AMERICAN PHILOSOPHICAL QUARTERLY

MONOGRAPH SERIES

WILBERFORCE UNIVERSITY
Wilberforce, Ohio
Learning Resources Center

AMERICAN PHILOSOPHICAL
QUARTERLY

MONOGRAPH SERIES

Edited by NICHOLAS RESCHER

DESCARTES'
PHILOSOPHY OF NATURE

JAMES COLLINS

Monograph No. 5   Oxford, 1971

PUBLISHED BY BASIL BLACKWELL
WITH THE COOPERATION OF THE UNIVERSITY OF PITTSBURGH

© *American Philosophical Quarterly 1971*
ISBN 0 631 11490 4

Library of Congress Catalog
Card No.: 77-149135

PRINTED IN ENGLAND
by C. Tinling & Co. Ltd., London and Prescot

CONTENTS

Preliminaries

IN René Descartes' famous metaphor of the tree of philosophical wisdom, metaphysics furnishes the roots, philosophy of nature the main trunk, and the practical disciplines the branches and ultimate fruits of wisdom. During the second half of the seventeenth century, his philosophy of nature received a lion's share of attention as an impressively argued and organized statement of the new mechanical view of the universe. But it was unable to prove its fruitfulness as a guide and framework for the physical research actually being conducted during these decades, due to its own lack of mathematical formalization and the wooden literalness with which the Cartesian school adhered to its imagery of the macroscopic and microscopic worlds. Every step in the spread of the Newtonian natural philosophy meant a further defeat and discrediting of Descartes' thought, especially in the area of philosophy of nature. Yeats' image of the dissolving center applies here with peculiar aptness: with the breakup of assent to his central trunk in the theory of nature, the entire philosophy of Descartes caved in and was transformed into dismembered fragments.

One lesson taught by the history of philosophy is, however, that the cultural defeat and dissolution of a philosophy are never entirely definitive. There are theoretical meanings and arguments which do not get entirely proportioned to a given cultural situation, and hence which have a lasting power which reaches beyond that situation and its dissolution. Using the resources of analytic and phenomenological methods, we are finding this to be the case with the philosophy of Descartes. Some of its aspects survive to reward our closest efforts at inspection and criticism. At least, the Cartesian theories of method, knowing, and being are once more coming under careful scrutiny and disclosing their capacity to illuminate the basic problems in philosophy. But far less work has been done in the broad area occupied by Descartes' theory of nature. It remains an open question whether anything of philosophical significance can still be salvaged here where the scientific inadequacies of the system were most incisively exposed.

The matter cannot be settled in a purely *a priori* fashion, without doing the preliminary historical explorations into what has become almost unknown intellectual territory. The recovery of Descartes' thought in the sphere of nature is a gradual process, being carried on by researchers in the history and philosophy of science and other disciplines. My aim in the present monograph is to contribute something to the historical understanding and analysis of the Cartesian conception of nature, insofar as that conception can be enlivened within the framework of his philosophical method and theory. Although this constitutes a restricted rather than an exhaustive approach to Descartes' vast speculations, it does seek to determine the general philosophical meanings of nature and their systematic role in his philosophy. In this way, his contribution to the persistent philosophical theme of nature can be weighed more precisely and can be made more accessible to our present theoretical interest in that theme.

To take account of the internal developments in Descartes himself, I have organized the investigation in two parts. The line of demarcation is drawn roughly at 1637, when he was finally able to publish the *Discourse on Method*. Prior to that time, he succeeded in determining the basic elements in his conception of nature, even though he had not worked out in detail the methodology, metaphysics, and general framework of his reasoning. Hence Part I treats of Descartes' theory of nature in its formative stage. But there is the further question of whether that theory requires any metaphysical grounding and receives any modifications by being brought within the general ordering of his philosophical inferences. Part II examines the process of incorporating the theory of nature within the Cartesian philosophy as a whole, where it serves as the unifying "trunk" of the wisdom tree.

This monograph is dedicated in friendship and respect to my longtime Jesuit colleagues in Philosophy: Robert J. Henle, S.J., George P. Klubertanz, S.J., Linus J. Thro, S.J., and William L. Wade, S.J. (1906-1968). The chairman of Saint Louis University's Department of Philosophy, Father Thro, has generously facilitated the preparation of the manuscript.

Part I
FORMATION OF THE THEORY
OF NATURE

Introduction

DESCARTES served a rigorous, self-imposed term of philosophical apprenticeship, which came to an end only with the publication of the *Discourse on Method* and its accompanying essays in 1637, at the age of forty-one. During this entire probationary and experimental period, he displayed a steady interest in the problem of nature. Whether he was engaged formally in mathematical and physical research or in drawing up the rules of method and sketching the outlines of metaphysics, he always managed to build a road leading from his direct topic to that of the meaning of nature. This practice indicates how central the question of nature was in his thought and how complex he found its actual investigation.

These early years of private study yielded Descartes two discoveries about the relationship between his mind and the meaning of nature. For one thing, he found a close relationship holding between the concept of nature presented in the Scholastic manuals on *philosophia naturalis* and the general framework of logical definitions, metaphysical principles, and physical methodology underlying these same manuals. In the degree that Descartes became dissatisfied with this general conceptual framework, he also became detached from the accepted views on nature. He could not call for a radical rethinking of problems in knowledge, metaphysics, and physical research without also committing himself to an equally profound reform of the very conception of nature.

Along with uncovering this reformation, however, he also discovered that he could make some definite progress in reminting the meaning of nature even before he himself had fully worked out the metaphysical bases and essential articulations of a new philosophical system. He did not have to wait until then, to be warranted to make fresh suggestions on the theme of nature. It was possible to establish a good deal about the method of studying nature, and about the content of its concept, at the same time that he was groping toward a new approach in the general methodology of philosophy and in metaphysics.

The task of Part I is to examine Descartes' thinking on nature during his crucially formative years. We will try to determine precisely how far he was able to conduct the reconsideration of nature under the intellectual condition of having some metaphysical ideas, but of not yet having worked out the new grounding of metaphysics and the practical aspects of his thought. In these sections, therefore, our focus will be upon the Cartesian theory of nature taken in an *as yet unincorporated* and unintegrated condition, or prior to its being systematically integrated into the philosophy-of-nature phase in the reconstruction of philosophical wisdom. This approach is required both by the fact that Descartes himself devoted so much attention to the problem of nature at this stage of his development and because of the permanent results which were assimilated to his mature position, but which are better understood in this genetic way.

1

A Fable for Skeptics

THE advantage of looking at the unincorporated Cartesian treatment of nature is that it permits us to examine some of the issues in a relatively simpler form and in sharper detail than is possible in the later system. One feature that immediately stands out in the early writings is Descartes' linguistic habit of coupling the term "nature" with some qualifying designation. He seldom permits himself to speak about nature in an absolute and unqualified manner, but refers to nature as being viewed under some specific aspect of the inquiring mind. Even when he deals with nature in general rather than with the nature of this or that thing, he is likely to state that he is treating nature precisely as seen within the framework of his new conception of the world.[1] It is not nature bare, but nature contexted by human methods of study and by determinate regions of reality, that comes within the scope of philosophical discourse.

Descartes carries over unobtrusively into his mature treatises this cautious linguistic usage in discoursing about nature. In refusing to talk about nature apart from some definite reference to the perspective concerned, he supplies a clue about his permanent motivations. His caution on this point indicates that his lifelong sensitivity to the skeptical challenge is specially lively on the meaning of nature. And it also furnishes to historians of philosophy a clear signal that the Cartesian project of reforming and reconstituting philosophical concepts demands an unusually close scrutiny of every shade of meaning intended for the term "nature."

The Greek skeptics were accustomed to give point and flavor to their attack on the savants and philosophers, who discoursed grand-

---

[1] Mechanical laws of nature are, more definitely considered, the rules "according to which we must think that God makes the-Nature-of-this-new-World act." *Le Monde*, chapter 7; in Descartes, *Oeuvres philosophiques*, ed. by Ferdinand Alquié (vols. I and II, Paris, 1963–1967), vol. I, p. 351. Wherever possible, references to Descartes' own Latin and French text will be made by volume and page to the Alquié edition. Descartes' primary targets, the qualitative and nonmechanical school philosophies of nature, are analyzed by P. Reif, "The Textbook Tradition in Natural Philosophy, 1600–1650," *Journal of the History of Ideas*, vol. 30 (1969), pp. 17–32. The life context of Descartes' scientific education is described by J. R. Vrooman, *René Descartes, A Biography* (New York, 1970), chapters 3 and 4: "A Revolution in Thought" and "Toward a Universal Science," pp. 68–134.

iosely about the properties of nature, by asking two simple questions: *Which nature are you talking about?* and *Does nature exist?*[2] Both these questions aroused a strong and constant resonance in Descartes' mind; he felt that he could not really fulfill the philosopher's vocation without trying to answer them. The first one reminded him that there is no universally comprehensive and accepted meaning of nature, despite the widespread scholastic practice of organizing every discussion around an analysis of the definition given by Aristotle. Any distinctive positions one may develop on basic scientific, philosophical, and practical issues are likely to involve a distinctive conception of nature itself. The second skeptical question warned Descartes that even the most finely executed and pluralistic conceptual reformulation is still open to the radical challenge of whether, and in what sense, the resultant meaning of nature concerns any reality in the existential order.

When Descartes' thoughts on nature are viewed (like so many other major themes in his philosophy) as prolonged responses to the skeptical challenge, then the division of labor governing his extensive studies becomes historically understandable. During his apprentice years he concentrated mainly upon clarification of the basic *meanings* of nature, without pushing the issue of their existential foundation to its ultimate philosophical determination. His chief concern was to specify the leading components required to give some definite significance to that concept of nature which is involved in the new mechanistic science. As a functional answer to the skeptic's first question, then, he sought to show which working meanings of nature would render more intelligible that view of the world being shaped for him by the new mathematical physics. While the question of *existence* troubled him from the outset and provoked some informal indications of where its resolution lay, he reserved the exacting philosophical treatment of the existential status of nature for his subsequent metaphysical inquiries.

Given his functional approach to the first skeptical question, it is not surprising that Descartes should devote considerable attention to the problem of nature precisely in the early work which sets forth

---

[2] In *Outlines of Pyrrhonism*, Bk. I, sect. 98, Sextus Empiricus sets forth the strategy of skeptical questioning on this issue. " 'But', one will say, 'nature made the senses commensurate with their objects'. In view of the great disagreements among dogmatists on the unsettled question of the existence of nature, we can only reply, 'Which nature?' " *Scepticism, Man, and God: Selections from the Major Writings of Sextus Empiricus*, tr. by S. G. Etheridge and ed. by P. Hallie (Middletown, 1964), p. 58. Modifications are made in the published translations used throughout the present monograph.

his mechanistic outlook: the *World*, along with its prolongation in his chapters on *Man*. These writings display his refusal to settle upon a reflective meaning for nature simply by making a choice among the traditional definitions catalogued in the natural philosophy textbooks or by attempting to conflate them into a compromise definition. Here as elsewhere, Descartes rejects the dilemma of being either a traditionalist or an exotic innovator. To avoid these extremes, he adopts the maxim of always determining his conception of nature by the most general theoretical requirements implicated in a scientific explanation of the sensible world. These requirements were being specified in his own age by a much more mathematico-physical use of human intelligence than the Greek philosophers of nature had achieved or than Telesio and the Italian philosophers of nature could match with their dreams. Thus to elaborate a theory of nature in conformity with the presuppositions of the mechanistic world view is the Cartesian path of liberation from the overladen history of the philosophies of nature, without becoming sidetracked into the slough of Renaissance magic and fantasy.

At the same time, Descartes sees that a certain price must be exacted for establishing a proportionate relationship between his general meaning for nature and the theory of the visible universe propounded in his treatise on *World*. He calls his account of the genesis and structure of the world a *tale or fable*,[3] and consequently he must assign a fabular status to his correlated theory of nature. What he offers is a fable for the skeptics and for all those contemporary minds who were sufficiently disturbed by skeptical queries to become fundamentally dissatisfied with the Aristotelian doctrine on nature. Yet his story of nature is not intended to be just another Renaissance substitute for Aristotelian nature, replete with appeals

---

[3] Descartes has "no other aim than to tell you a tale." *Le Monde*, chap. 7 (Alquié, vol. I, p. 364). Writing to Mersenne, he refers to this work as "the fable of my World." *Letter of November 25, 1630* (Alquié, vol. I, p. 285). To focus on problems surrounding nature, I use my own translation of Descartes' letters. But one should consult Descartes, *Philosophical Letters*, translated by Anthony Kenny (New York, 1970). (Cited hereafter as the "K" translation; see p. 18.) The term "tale" or "fable" has no pejorative meaning in Descartes' early physical theorizing, but underlines the circumstance that such theorizing is not yet incorporated within a framework of philosophical principles. By 1645, however, Descartes sometimes uses the term "tale" or "fable" as the equivalent of a sheerly fabricated hypothesis, contrasting it with those hypotheses which do rest upon basic metaphysical and mechanical principles and which do seek verification in the phenomena of nature. See *Letter of May, 1645*, to Mesland; in *Correspondance de Descartes*, ed. by C. Adam and G. Milhaud (8 vols., Paris, 1936–1963), vol. VI, p. 236. Descartes' scientific interests are described by J. F. Scott, *The Scientific Work of René Descartes* (London, 1952).

to occult qualities and magical forces. His is an intellectually responsible tale, both because it rests on analysis of some general
requirements involved in the scientific account of things, and because
it accepts the limitations inherent in applying the machine analogy
to nature.

There are some specific reasons why Descartes designates his theory
as a *fabula*. A strongly prudential consideration is present, of course,
lest his own theory of the world suffer the same repressive fate as
befell that of Galileo. Descartes is concerned not only for the future
reception of his own thoughts but also for the credibility of the
Church's pronouncements on the universe. When unsophisticated
churchmen intervene in discussions on the antipodes or on planetary
motions, they betray the lack of a methodology for differentiating
between religious, scientific, and philosophical ways of viewing
nature. By attaching some qualifications to the reference of his theory
of nature to reality, then, Descartes seeks to forestall the category
mistakes made in the case of Galileo.

Yet he also recognizes some intrinsic epistemological grounds for
taking a cautious attitude toward every account of nature, including
his own. The treatise on the *World* opens with an incisive and quite
general critique of the reliability of the senses and the naive pre-
sumption that things must resemble the qualitative perceptions and
images we have of them. In particular, Descartes proposes that the
nature of light as an organizing principle of the world be treated in
distinction from the question of human perception of light. Under
such conditions, we have to suspend our assent to all previous
philosophies of nature. They rest upon an uncritical attitude toward
sense perception and the relation between our imagery and reality;
moreover, their insensitivity to current scientific research deprives
them of any internal principle for reforming their conception of
nature along more fruitful lines. By invoking the need for a skeptical
interval in our reflections on nature, Descartes clears the ground for
presenting his own view as a new tale to be judged independently by
its own coherence and its agreement with a scientific study of light
and mechanical relations.

There are some other circumstances in the telling of his tale about
the visible world which underline the fabular, yet disciplined,
character of his conception of nature. Although seeking to eliminate
the fantastic and magical touches of a Paracelsus or even a Francis
Bacon, Descartes does admit and indeed stress the need for a well-
controlled use of imagination. We must employ it along with critical
intelligence in order to frame a likely story concerning the primordial

formation of the world from chaos, as well as to depict the mechanical motions and relations among the ultimate particles of material things. Insofar as the Cartesian notion of nature incorporates the imaginative use of the mind in suggesting a mode of origin and developing mechanical models for understanding the present structures in the world, this notion retains the fabular traits of the entire mechanical explanation of the world.

In addition, the Cartesian treatment of nature in the writing on *World* is subject to two radical limitations. (1) That treatise does not furnish the philosophical resources for ascertaining the relationship between the mechanical explanation of the world and the existential order of reality. In consequence, the theory of nature remains itself an unfinished tale, one which does not yet face up to the skeptical question of existence and hence which fails to settle its own bearing upon existing things.

(2) As we will see presently in discussing eternal truths, Descartes assigns a role to God in the ordering of the universe, even considered as having a mechanical genesis and structure. But the *World* employs a set of presuppositions about God, without being able to supply the philosophical basis for accepting them. Since these theistic presuppositions figure quite prominently in the Cartesian conception of nature, the latter is proposed to students only as a well-conceived but incompletely established story. This signifies that the view of nature is not yet fully integrated with its required philosophical bases. At this stage, Descartes' tale of nature remains weaker than a strict hypothesis in his philosophy of nature, since the latter would provide an adequate context of metaphysical principles and theistic inferences in the existential order.

2

Demythologizing the Eternal Truths

THE incompleteness of the Cartesian tale of nature is strikingly exhibited by one circumstance of its telling: its constant use of a reformed meaning for the eternal truths. There is no obvious connection between these two topics, since the problem of the eternal truths belongs formally within Descartes' metaphysics. But when his early works are studied genetically from the metaphysical viewpoint, they reveal a surprising imbalance of treatment. For although his main metaphysical positions are only lightly sketched before 1641, his internal stand on the question of the eternal truths is worked out in considerable detail and is communicated to influential correspondents. A major reason for this lopsided development of one metaphysical theme is found in its relationship with the theory of nature. Descartes needs to have at his disposal a fairly well-elaborated conception of eternal truths, even before he is in a position to establish all the metaphysical presuppositions of his argument. Although the incompleteness of the metaphysical founding for his early notion of eternal truths is communicated to his theory of nature as well, the co-implication of the two topics is too close to permit him any leisurely postponement of the analysis of eternal truths until the proper moment within the order of metaphysical reasoning.

What interests Descartes the natural philosopher in this issue is its involvement in many currently attractive views of nature as an independent, all-determining entity, into whose toils man is totally swept up. Whether this picture of nature is rendered in the style of Pomponazzi, Bruno, or even Galileo, it supposes that the sphere of natural processes is regulated by inexorable laws having an independent binding power over both human and divine agencies. Proof for the autonomy and irresistible force of these laws of nature is sought in their being an expression of independent eternal truths. God himself is conceived as being bound to acknowledge such eternal truths and conform his operations in the world to the exigencies of the natural processes governed by these operative truths, or laws of nature.

In Descartes' estimate of the situation, however, men are permitting themselves to be overawed by an idol of their own making, yet one to which even the divine power is being asked to submit.

This idol (or, in more recent language, alienational projection) is not located precisely in physical nature, but rather in man's justifying notion of the eternal truths shaping the laws of nature from within. To demythologize this notion is a primary critical task in the Cartesian theory of nature, since it stands in the way of a thorough reconstruction of the meaning of nature. One way to *de-autonomize the eternal truths* is to insist upon their dependence on a twofold context in the creative mind of God and the judging mind of man. Hence from the outset, Descartes is prepared to employ some metaphysical considerations about the divine and the human minds, even though he is not yet in a position to substantiate them in full.

He outlines his basic strategy in correspondence with his friend and intellectual broker, Marin Mersenne.

> I will not fail to touch in my physics on several metaphysical questions, and particularly this: that mathematical truths, which you call eternal, have been established by God and depend on him just as entirely as all the rest of the creatures. To say that these truths are independent of him is, in effect, to speak of God as of a Jupiter or Saturn, and to subject him to the Styx and the Fates. Never fear, I beg you, to assure and proclaim everywhere that it is God who has established these laws in nature, just as a king establishes laws in his kingdom. . . . I do not conceive of them as emanating from God, like rays from the sun. For I know that God is author of all things, and that these truths are some thing, and hence that he is their author. . . . From the very fact that he wills some thing, he therefore knows it, and for this reason alone such a thing is true. Hence one must not say that if God did not exist, nevertheless these truths would be true. For God's existence is the first and most eternal of all the truths that can be, and the sole one whence all the others proceed.[4]

As this composite passage indicates, Descartes is aware of his predicament in having to use a detached metaphysical position on the meaning of the so-called eternal truths as a necessary initial step in

---

[4] *Letters of April 15, May 6, and May 27, 1630*, to Mersenne (Alquié, vol. I, pp. 259–260, 264–265, 267); selected portions of these letters are translated in Descartes, *Philosophical Writings*, tr. by E. Anscombe and P. Geach (New York, 1954), pp. 259–263 (cited hereafter as the "AG" translation). By contrast with the uncreated eternal truth of God's nature and existence, all created eternal truths "are something less and something subject to this incomprehensible [divine] power," in which respect they are less eternal. *Letter of May 6, 1630* (Alquié, vol. I, p. 265). On the paradoxical Cartesian expressions: "more and less eternal," "more and less contradictory," "more and less *a priori*," consult H. Gouhier, *La Pensée métaphysique de Descartes* (Paris, 1962), pp. 233–264, 285–291. Their function is to point up God's uniqueness as the creative source—the *fons veritatis* in the order of being—for all our truths about nature and man. The relevant texts are given in Descartes, *Philosophical Letters* (K, pp. 8–16).

his theory of nature, even before he is in possession of a fully argued metaphysics of God and creation. No further progress can be made in rethinking the concept of nature, however, until the claim of independence for eternal truths is removed and their quite creaturely or mind-dependent condition established.

Descartes consistently reminds Mersenne and others that, in talking about eternal truths, they are uncritically following a current conventional usage. The first step in liberating oneself from the tyranny of such usage is to notice the very capacious scope of what are called the eternal truths, and to specify the kind of referent intended in each instance. There are more referents included than the axioms and other basic propositions of pure mathematics. Descartes uses the language of eternal truths in a broader denotation to include at least three other conceptual regions. They are: the general mathematical concepts embodied in the mechanical laws of nature formulated by Galileo, Mersenne, and Descartes himself; the metaphysical axioms or common notions involved in thinking about nature in general and particular natures; and the simple natures or basic units of meaning in which the Cartesian analysis of thinking and extended reality terminates, and out of which the construction of the Cartesian visible world is made. Because of this broad range of talk about the eternal truths, reaching into our chief resources for investigating nature, there is no standard and wholly unambiguous way of stating the meaning of nature prior to a philosophical analysis of the usage.

One move can be taken by determining the import of Descartes' remark that "these truths are some thing." They fall within the order of things, only when the term *thing* (*res, chose*) is used in the very broad sense of designating all the objects of human perception or cognition. Descartes usually divides the objects of our perception into three groups: existing beings (things in the most proper meaning), affections or qualities of existing things, and the eternal truths. The noteworthy point here is that only the first two classes of objects of human cognition are existential in their own direct signification. Yet no matter which of the above mentioned referents for eternal truths is intended, it does not signify by itself anything in the existential order. To say that an eternal truth is some thing, in the broadest designation, is only to affirm its function as an object of our perception. It is a reality, in the sense of constituting a unified meaning for our mind, but this is not sufficient to establish for the eternal truth any direct and autonomous hold upon existing reality.

The existential reference of eternal truths involved in the interpretation of nature is achieved only through their relationships with the minds of man and God. They have the common characteristic of being known by our finite mind in such a way as to be *comprehended* by it. Hence they are proportionate to the finite reality of the comprehending human mind, however permanent and universal their truth may be. This proportionate relationship assures Descartes of two conclusions about eternal truths which run counter to the myth of their being independent structural laws for everything. (1) Their status of reality or thinghood is not that of a self-grounded causal dynamism, but consists precisely in their modality as objects of human perception and instruments of inquiry into nature. (2) And because their modal reality is comprehended by the human mind, it is finite and dependent entirely upon the free, creative act of the divine mind.

Thus both the human and the divine poles of the existential reference of eternal truths show that the latter enjoy no subsistence and causal determination of their own. By working out their proportion to the comprehending human mind and their total dependence upon the creating divine mind, Descartes weakens the naturalistic appeal in his age to the eternal truths. Once we analyze their type of reality, we see that it cannot be used to confer an unconditionally independent status upon the laws of nature and the natural process as a whole.

To reinforce the divorcement between eternal truths and an eternalistic naturalism, Descartes often makes an instructively paradoxical comparison between truths that are less eternal, more eternal, and most eternal. This unusual usage is deliberately intended to make his contemporaries more wary about calling our axiomatic principles and basic concepts eternal in an unqualified way. If we look at their status in reality by comparison with the divine reality itself, we must denominate them as "less eternal" truths. Their eternality is of a conditioned and established sort, being wholly the outcome of the divine act of creative knowing-and-willing, which does not require their anterior presence in order to specify itself. Our human axioms are eternal only in the restricted sense that they are established by God to hold good unchangingly for our actual world and any world similarly structured. Their universal validity and immutability remain completely derived and dependent, as can be seen when skeptical questions are raised concerning the basis for claiming their relevance for man's study of nature.

In striking contrast are what Descartes terms the "more eternal" truths of the divine being and power taken in themselves. They constitute the realm of the underived and radically originative truths, those which signify the foundational reality of the creative God. Their eternality is indeed autonomous and unconditional, since their significance is to express a reality unestablished by any other agency and also the powerful creative source for all other truths and laws. We can know something about the unconditionally eternal truths but only in a participative manner. What we can never do is to comprehend them perfectly and thus render them proportionate to our finite minds and purposes. Hence they always remain the *more* eternal truths, those which we acknowledge and yet cannot reduce to human stature and full comprehension. The aspect of transcendence involved in every inquiry about man and nature has its roots in the foundational relationship of the more eternal truths to our human mind and our axiomatic truths.

Finally, in the text upon which we are commenting, Descartes refers to God's existence as "the first and most eternal of all the truths that can be, and the sole one whence all the others proceed." The use of the superlative at this point is due not to pious rhetoric, but to the twofold function which the truth of God's existence must serve in Cartesian thought. Internal to the divine reality itself, this truth emphasizes the self-dynamism and active power of the divine essence in founding its own existing actuality. And in respect to everything else, including the reality of the less eternal truths underlying the human sciences, this truth measures their utter dependence for a real foundation upon the creating power of the divine mind and will.

Descartes can now drive home his point against the existential autonomizing of the eternal order of nature. When an existential significance is sought for those eternal truths governing our knowledge of nature, our mind must submit such eternal truths to the prior conditions set by the existing active reality of God, rather than submit the divine activity to such truths. We cannot think truly about the existential bearing of our concept of nature and its axiomatic framework, unless we adjust the latter to the mediation of the fundamental truth of God's existence. In the *reversal of perspectives* required by this theistic mediation of all existential assertions about the reality of nature lies Descartes' chief weapon for overcoming the naturalistic tendency to absolutize the order of nature.

Through his comparative approach to the theory of eternal truths, he has equipped himself with a decisive context for interpreting all references to the great machine of nature, the laws of nature, and the order instituted by nature. Descartes feels free to use such phrases, since they remain qualified in principle by the limits set for the less eternal truths falling within human comprehension. Once the philosopher of nature is alerted to the distinctions required in careful talk about the eternal truths, he has a corrective against giving direct existential value to the image of a wholly self-sustaining machine of cosmic process. He realizes that the laws we formulate for nature come within the range of the comprehended sort of eternal truths, and hence that they are really subject to a mediational reference to the free creative act of God. Descartes' advice is that we spell out this reference sometimes by explicitly regarding the order of nature as constituted by "certain laws which God has established in nature."[5] Qualified in this way as a result of the critique of eternal truths, the concept of nature can now be developed in a perspective that is at once mechanistic and theistic.

[5] *Discourse on Method*, Part 5; in Descartes, *Discourse on Method, Optics, Geometry, and Meteorology*, tr. by P. J. Olscamp (Indianapolis, 1965), p. 34. E. Gilson, *René Descartes, Discours de la méthode: Texte et commentaire* (2nd ed., Paris, 1939), pp. 372–373, remarks that "established" must be taken here in the strong sense of a creative imparting of a pattern or action to matter for the determination of physical phenomena. See É. Bréhier, "The Creation of the Eternal Truths in Descartes's System," in *Descartes*, ed. by Willis Doney (New York, 1967), pp. 192–208.

3

Three Primary Meanings of Nature

DESCARTES distinguishes between two major tasks facing the philosopher of nature: he must make a quite general analysis of the meaning of nature, and he must also make a progressively specified study of the several kinds of natures found in the world. Our present concern is with the most general level of investigation, since the more particularized approach to natures raises the question of special methodic aids of explanation which will engage us in the latter half of this study. In making a general consideration of nature, one seeks to determine those most basic requirements for describing and explaining the structure of the perceivable world and any other world that could show a similar organization of phenomena. Descartes identifies three such requirements, which thereby establish their right to comprise the three primary meanings of nature. They are: (a) God the creator and conserver, (b) the material aspect or concrete extension, and (c) the formal aspect or laws of movement.[6] It is out of the interwoven relationships between these principles that the general Cartesian significance for nature is born.

Operationally stated, to develop a general theory of nature means to look upon the world (whether our actual one or a fabular reconstruction) as being jointly constituted by the divine power and the coordination between extensive material particles and the pattern of mechanical movements. This operational rule of correlating three distinct considerations of nature is Descartes' safeguard against the monistic attempt to fuse all aspects of nature into a single reality and

---

[6] *Le Monde*, chap. 7 (Alquié, vol. I, pp. 349–351). See the comparative texts given in E. Gilson, *Index scolastico-cartésien* (Paris, 1912), s.v. "nature,' pp. 197–202. For the Cartesian use of the expression *Dieu ou la nature, Deus aut natura*, see *Letter of April or May, 1638*, to Reneri for Pollot (Alquié, vol. II, p. 56); and *Principles of Philosophy*, Bk. I, sect. 28, in *The Philosophical Works of Descartes*, tr. by E. Haldane and G. Ross (2 vols., New York, 1955), vol. I, p. 230 (cited hereafter as the "HR" translation). The medieval background for regarding nature as: (i) the creative *natura naturans* of God, (ii) a "goddess" or theophany of divine power, and (iii) the desacralized realm of causal process and scientific inquiry and even human technology, is furnished by M.-D. Chenu, *Nature, Man, and Society in the Twelfth Century* (Chicago, 1968), pp. 1–48. Descartes' speculations about "automata made by God or nature" (*Philosophical Letters*, K, p. 54) join meanings (i) and (ii), but at the expense of meaning (iii).

agency. It cannot be inferred from a *unified theory* of nature that the *reality signified* by the theory is a massive whole of being, whether conceived as an existing entity interposed between God and man or as a totality embracing them both in the single reality of nature. What the Renaissance Stoics and the Averroists failed to notice was that the complex signification of nature depends on an intellectual synthesis between several real principles and several interpretative acts of the human mind. The Cartesian emphasis upon the three primary meanings of nature is designed to render us critically aware of the pluralistic conditions in reality and thought which render possible a unified theory of nature, but which are never revoked by the complex signification thus achieved.

Like his medieval Christian and Jewish antecedents, as well as the authors of current Scholastic manuals in natural philosophy, Descartes often refers to *Deus aut natura—God or nature*—without fearing any pantheistic overtones. Sometimes, he employs this first basic meaning of nature in order to underline the genetic unity of the world. All the material particles, movements, and varied appearances take their real origin from a single divine source, which thus assures their relatedness and tendency to constitute a universe. Another function for God, considered as a primary significance of nature, is to bridge the distance between the elementary particles of extended matter and general patterns of mechanical motion and the very complex and diversified world of our actual experience. By including the infinitely active God within the principal meanings of nature, Descartes suggests that the inexhaustible power and fertility of the divine reality are proportionately communicated to the extensive and mechanical components in nature.

Thus the theistic aspect of the Cartesian tale of nature corresponds to the human mind's strong need to find an adequate foundation for the relatedness and unity, the fertile variation and complexity, of our natural world. Because God is included in the general significance of nature precisely in respect to his infinite and unifying causal power, he is not identified in being with the material constituents and the modes of motion themselves. The originative manner of God's inclusion in the meaning of nature prevents any ontological fusion of his reality with that of the material, moving world.

Nevertheless, Descartes perceives that the traditional path of reasoning from the visible world to God is vulnerable both to skeptical criticism and to his own doubt about the foundational validity of sense testimony. Until he can develop a metaphysical way of inferring God from his own doubt-resisting reality as a thinking being, and not from the order of sensible things, he must remain content with the common conviction of men of faith in the divine creator. It is from this consensus of religious theism, rather than from any direct argumentation for the reality of God, that the meaning of God as nature draws its living assent during the first telling of the Cartesian tale of nature. Thus there is a notably provisional and pistic quality about the grounds upon which Descartes rests his view of God as nature. He is under obligation eventually to furnish a metaphysical basis for accepting this signification, along with a philosophical order of inferences within which to regulate its function for the philosophy of nature.

Supposing the truth of the theistic conception of God for the sake of his preliminary inquiry, however, Descartes' next step is to specify something about the transition from the divine source of nature to the material and formal meanings of nature in general. This is not a metaphysical question about the creative act itself, but rather asks about the original condition of matter and its adaptation for the mechanical laws of movement. In the course of determining the second basic meaning of nature in terms of its *material component*, Descartes has perhaps his most serious confrontation with the prevailing religious conception of things. For although he depends upon this source for general theistic assurances, he does not hesitate to reform the religious outlook prudently to bring it into closer conformity with all aspects of his own account of the genesis of nature.

To avoid ecclesiastical condemnation on the same scale as that in the Galileo case, Descartes did not hesitate to wrap his particular teaching on the earth's motion in the elaborate casing of his theory of vortices. But the very generality of the problem of how to understand nature in its original material condition did not permit any evasion or any reformulation within a wider frame of reference. In determining the second general meaning of nature as material particles, therefore, Descartes was unavoidably required to make a distinction between the basic theistic assent itself and some features in the popular religious and theological thinking on how the world proceeds from the hand of God.

Specifically, he could not accept the solidarity between believing that things come from a supremely intelligent and powerful maker and stipulating that they must display *at once* the structural order of our presently perceivable universe. Descartes was forced to challenge the assumption that God the free creator must bring forth an already harmonious world, one which exhibits from the outset those traits of variety and relationship now familiar to us. Had he not made a reform of the prevalent religious outlook at this point, he could not have developed his own position on the material component of nature and its openness to mechanical motions.

Yet the Cartesian critique did not bite any deeper than was necessary for clearing a path for an explanation of nature in terms of extensive particles and mechanical movements. Descartes did not transcend his own age and its interpretation of the Bible story as assigning an age of only a few thousand years to our universe. He shared neither in Giambattista Vico's sense of nature as a constant living growth nor in our own evolutionary perspective of time measurements. He did have some relative appreciation of the time factor in a particulate-mechanical explanation of nature, but his use of this factor remained quite indirect and secondary. Cartesian physical time is mainly a function of the constantly operating mechanical laws in respect to the spatial disposition of bodily particles. There must be at least a sufficient passage of mechanically conceived and spatially measured time to secure the operation of the laws of motion upon material particles.

The notion of time as a lapse, permitting mechanical operations to occur, is Descartes' way of rendering more credible the interplay between his second and third meanings of nature. Thus he eliminates the need for occult forces to account for the present constitution of bodies and their qualities. He also recognizes the importance of the time factor, in the degree that his mechanistic explanation of nature includes some reference to the wear-and-tear of materials already well formed and the smoothing-out of particles in the course of their constant interaction.

There is another indirect role for time, once the agency of man is introduced into the tale of nature. At this juncture, Descartes admits the temporal difference made in our control over natural processes by the increase of human experience on the part of artisans, as well as by the increase of direct knowledge of the general and particular meanings of nature by the theoretician. The historical developments in technology and physical theory cannot fail to affect the meaning of nature, even within Descartes' weakly historical perspective. Yet there is no primary temporal principle shaping the Cartesian interpretation of nature: time never becomes a crucial and creative element in its connotation.

Within his pre-evolutionary setting, Descartes expends considerable energy rather in a lifelong struggle to reconcile the current understanding of the Biblical story of the world's genesis with his own scientific account.[7] He never matches the supple epistemological and methodological distinctions proposed in Galileo's *Letter to the Grand Duchess Christina* concerning the relationship between the Bible, theology, and scientific theories of the world. But he does provide himself with some guidelines for treating the conflicts that arise for anyone consulting these three sources of interpretation of nature. Even though Descartes' views on the problem developed slowly across his entire lifetime, I will group his three main points conveniently together here.

(a) His early position on this issue was tentatively advanced and then gradually modified, in proportion as his confidence in the principles of his own philosophy increased. At first, he thought it sufficient to insist in Galilean fashion that his own scientific and philosophical hypotheses need not be taken as describing the actual processes of the real world. The Cartesian tale of the world in its origin and differentiation states the explanatory conditions required for a mechanical model of nature, but it need not necessarily specify the real causes shaping the development of the actual world. Thus a difference of viewpoint can legitimately arise between the Cartesian principles of nature which facilitate a mechanical explanation (interpreting the initial conditions in terms of nature as an in-determinate matter or chaos), and the current theological assumptions (interpreting the world in terms of a determinate harmonious order from the very outset).

But this easy irenic solution of the differences proved unsatis-factory, in the long run, to Descartes himself. The more he insisted that the mechanical principles used in his tale of the world would hold good in *any* world manifesting the same appearances as ours, the more did he come to see that his genetic account in terms of the material meaning of nature should also characterize the *real* genesis of natural things. Hence as the distance began to close between Descartes' early tale of nature and his metaphysical analysis of the principles of the actual world, he proposed two further distinctions for reconciling his own theory of the world's formation from an

---

[7] *Le Monde*, chap. 6 (Alquié, vol. I, pp. 343–347); *Discourse on Method*, Pt. 5 (Olscamp, p. 37); *Principles of Philosophy*, Bk. III, sects. 45–47 (AG, pp. 224–226). Descartes defends the philosophical theory of a slow, gradual genesis from seeds or primal undeveloped conditions, on the epistemological grounds that it aids us more in determining and communicating the meaning of nature than does the postulate of a fully formed universe. Cartesian knowing is not quite so atemporal and insensitive to development, when there is question of the formation of physical nature, as M. Grene maintains in *The Knower and the Known* (New York, 1966), pp. 88–91. See the section on Cartesian cosmic evolution in F. C. Haber, *The Age of the World: Moses to Darwin* (Baltimore, 1959), pp. 60–98.

original chaotic matter with the conventional theological reading of Genesis in his day.

(b) One move was to leave this theological view standing, but to confine it to some presumed condition of things prior to the Biblical flood. Because of the link between laws of nature and the dependence of eternal truths on the divine freedom and power, Descartes could accept the abstract possibility that the constitution of nature may have been different before the flood than after it, assuming that the flood was universal and cataclysmic in its effect. On this assumption, he could then point out that the intent of his philosophical theory of nature is to determine the constitutive principles of nature as it now is in our experience (*natura prout jam est*).[8] Once the Cartesian tale of nature is incorporated within the metaphysically founded philosophy of nature, it does specify the primary meanings of presently experienced natural reality, whatever the conditions before the flood may have been. Yet this concession to the contingency of nature's constitution from the divine standpoint did not really lessen the universal scope of Cartesian explanatory principles of nature. They could be consistently construed as applying also to the world described in the Bible, including the flood story itself.

(c) Descartes' final act in this very sensitive area was thus to observe that, after all, the Bible is phrased in metaphorical language and adapted to the popular imagination in a pre-scientific age. A fresh approach to the primary meanings of nature and its genesis must be expected, in the degree that the inquiring mind succeeds in disciplining itself through mathematics and mechanics. Since these were precisely the instruments guiding his own conception of nature, Descartes suggested that his philosophical account could effectively help contemporary men of faith to reinterpret the Biblical metaphors on the formation of things. This suggestion was in line with his general conviction that religious faith benefits more from a free alliance with the Cartesian mechanical philosophy of nature than

---

[8] In defense of a distinctively philosophical approach to nature's formation and structure, Descartes argues thus: "It could be that, before the flood, there was another constitution of nature which was rendered worse by it [the flood. But] the philosopher considers nature (and also man) solely as it now is, and does not investigate its causes any further, since anything further surpasses him." *Entretien avec Burman*, ed. by C. Adam (Paris, 1937), p. 126. What surpasses the philosopher is not the idea of past development in nature (an idea which he can indeed reach from analysis of nature and man in their present condition), but rather the revelational basis for affirming with certitude that such development has actually occurred. On the metaphorical, humanly accommodated presentation in *Genesis*, see *ibid.*, p. 92, and *Letter of June 6, 1647*, to Chanut (K, pp. 220–225).

from theological reliance upon the Aristotelian schema of hierarchical forms and qualities in nature.

Descartes has a problem intrinsic to his own approach, however, of distinguishing his second primary meaning of nature from a sheerly poetic fancy of the Greek myths about darkling chaos. He meets the difficulty by always treating the material principle of nature in the context of a dual relationship: to the divine creative power and to the pattern of mechanical movements. Once we accept the derivation of the material component in nature through a free creative act, we must also admit that the human mind cannot use God's creative freedom as a sole premiss for making necessary deductions about the meaning of matter. As far as human inquiry is concerned, the more determinate meaning of material nature depends upon exploring the proportion between the material factor and the laws of mechanical movement which do come comprehensively within our range.

Thus nature in its second primary meaning signifies that proper subject in which the mechanically analyzable motions are received and communicated. In the broadest sense, the material component in nature must be proportionately *such* a reality as can receive the pattern of movements regulated by the laws of mechanics. What differentiates the Cartesian conception of nature-as-matter from the imagery of a primordial chaos, then, is just this requirement of its being intrinsically prepared for the reception of the movements studied in the Galilean science of dynamics.

As entering into the general significance of nature, Cartesian matter is neither a Platonic featureless receptacle nor an Aristotelian pure potency. Instead, it has a sufficiently determinate reality *to meet a set of immanent suppositions* correlating it with the mechanical motions.[9] In general, the matter in question must be of such a makeup that the movements it receives are mechanically describable and that its structure is picturable by means of sensible models, which can be related to the organized phenomena in the world. This frame of

---

[9] *Le Monde*, chaps. 6, 7, and 10 (Alquié, vol. I, pp. 346, 350–351, 357, 365); *Meteorology*, Discourse 1 (Olscamp, pp. 264–265); *Principles of Philosophy*, Bk. III, sect. 46 (AG, p. 225). Because this context of suppositions qualifies so deeply his entire theory of components in nature, Descartes feels obliged to use the very guarded wording quoted in the next paragraph below, when treating of extension as the essential attribute of matter in *Principles of Philosophy*, Bk. II, sect. 64 (AG, p. 221). He always emphasizes the suppositional status of, and verificational demand upon, the more particularized principles in his explanation of physical phenomena. "I would not dare affirm that the things I announce are the true principles of nature, but at least I will say that, in taking them as principles, I am accustomed to satisfy myself about all things which depend on them." *Fragment of 1633–1635* (Alquié, vol. I, p. 486).

reference enables Descartes to make some definite predications about matter and so assign it a controllable function in the general conception of nature.

His basic supposition about nature materially considered is that it consists essentially in extension. Descartes does not thereby sheerly equate the being of matter with extension, but designates the latter as the essential attribute, or that which brings material nature within the scope of our mind. This functional relationship between predicating extension as the essential attribute of matter and securing the knowability of material nature underlies the deliberately performative vocabulary which Descartes employs to describe the supposition. His focus upon extension is a way for him to "treat" and "consider" matter. A study of extension as the key trait yields some definite results, which he regards as "counting" decisively for the project of specifying nature's meaning in its material aspect. Such expressions manifest the epistemological aim of the attribution of extension, so that the general material meaning of nature can be clarified by reference to human science and activity.

The extendedness fulfilling this role is not pure geometrical quantity but concrete quantity. Extension is concretized first by adapting the basic geometrical concepts to the system of mechanics, and next by making passage from abstract mechanics to the mechanical interpretation of bodily motions in the world. In the course of this concretizing process. Descartes anticipates Leibniz's objection that, since extension is essentially inert, it lacks the capacity even to receive and sustain the mechanical motions found in bodies. Abstract extension, considered by itself as an isolated concept, would lack this capacity. But the concretization of extension means the consideration of it as an *interrelated reality*, not as an isolated concept. The adaptive process establishes its functional meaning both in respect to the mechanical analysis of bodies, and in respect to the divine power, since the general meaning of nature-as-matter arises only from making this twofold reference to mechanical laws and divine power (the other two primary senses of nature). Matter as concrete extension is thoroughly open to the reception of movement in accord with mechanical laws, because it is a reality open to the divine creative source of all kinds of movement and agency. The receptivity of matter to mechanical motions expresses the mutual adaptation between the primary meanings of nature in its material and formal aspects.

More is required than concrete extension and its general adaptiveness to mechanical motions, however, in order to have a fully functional conception of the material principle in nature. Descartes customarily spells out a number of more specific suppositions, whose aim is to assure that mechanical motions will be received into precisely that sort of material subject which can then be worked up into recognizable bodily configurations. Moving far away from the Aristotelian notion of a purely potential principle, he requires the material constituent in nature to consist of particles that are plural in number, diverse in size, shape, and texture, and thus open to differing velocity and direction of motions. Along with many scientific researchers and philosophers of nature in the seventeenth century, therefore, Descartes proposes a particulate and corpuscular conception of matter as a general component of nature. This conception of nature can also be characterized as mechanical, as soon as the motions of which the material particles are receptive are specified as being those regulated by the basic laws of mechanics.

In maintaining a particulate meaning of matter, nevertheless, Descartes does not demand that it be allied with a strict atomism. Material particles need not be taken as being indivisible in principle, since a reconstruction of the world of mechanical motions can be made without requiring such a stipulation. Hence Descartes regards the atomism of Gassendi as a piece of needless dogmatism, incapable of being verified and quite expendable for the work of understanding nature.

The cautious Cartesian meaning of the material component in nature refuses to place *a priori* limits upon either the power of God over material particles, or the progress of human instruments, devised for ever deeper penetration of the structure of such particles. It is sufficient for this theory of nature to offer a *middle-sized* determination of the material subject of motions as being comprised of particles that are divisible, many, and diverse—just provided that the particulate properties of size and thickness, shape and velocity, are recognized as coming within the range of geometry and mechanics. For these particulate suppositions about matter guarantee some access of the human mind to the material aspect of nature. The intelligibility of Cartesian nature-as-matter is assured by its contextual reference both to God as creator and regulator of mechanical movements and to the human sciences directly analyzing these movements.

That is why, in treating the material aspect of nature, we have already prepared the soil for the third primary meaning of nature. Nature in its general formal aspect does not refer to some universal substantial form or a cosmic soul, which metaphysical notions

Descartes explicitly rules out. He restricts the responsible meaning of *nature-as-form* to the mechanically regulated pattern of movements undergone by the material particles. In determining the mechanical rules of motion, one is also settling upon the most general and humanly ascertainable sense of the formal aspect of nature. Hence Descartes identifies these mechanical rules of motion with the laws of nature, taken in its formal signification. His circumstantial treatment of the laws of mechanical movement is his characteristic method for giving some determinate content to the third primary meaning of nature as form.

4

Laws of Nature

IN an illuminating text, Descartes reaches a meaning for the laws of nature through the same argument which establishes that there must be a general formal aspect of nature, in addition to God and the material aspect.

Know that by nature I do not understand some goddess or some other sort of imaginary power. I employ this word to signify matter itself, insofar as I consider it with all the qualities which I have attributed to it, taken together, and under this condition that God continues to conserve it in the same way that he has created it. For solely from the fact that he continues thus to conserve it, it follows with necessity that it [matter thus qualified] must have several changings of its parts, which [changings] not being able (it seems to me) to be properly attributed to God's action, since the latter does not change, I attribute them to nature. And the rules according to which these changings are made, I call the laws of nature.[10]

Here, Descartes seeks to prevent the apotheosis of nature in its particulate constitution, not only by noting its essential need for continuous divine conservation but also by showing that there must be a formal principle of change intrinsic to the reality of nature, but distinct from God himself and the particles of matter.

If the divine act of creation-conservation of movement in general were alone required in our conception of nature, then we could never explain the many diversifying mechanical movements of bodies. For God's immutability would be best expressed by motion in a straight line and in a perpetually undiversified manner. Descartes maintains this position against the ancient bias in favor of circular motion as a divine image. In fact, however, the world contains many kinds and shifting modifications of curvilinear motions, for the examination of which men have had to develop the geometrical and mechanical sciences. It is necessary, but not sufficient, to suppose that there are different quantitative dispositions among the particles of matter for diversifying the basic fund of movement, which supposition is compendiously stated in the meaning of nature-as-matter. What Des-

[10] *Le Monde*, chap. 7 (Alquié, vol. I, pp. 349–350). On the close adaptation between the initial chaos, the laws of movement, and the differentiation of particles, cf. N. K. Smith, *New Studies in the Philosophy of Descartes* (London, 1952), pp. 115–118.

cartes must now add explicitly is that the differentiating process consists of some mechanically describable general modes of change, leading to the distinct formal meaning of nature.

In the fully articulated Cartesian sense, the general formal aspect of nature gets expressed in the mechanical rules governing the movements whereby bodies are formed and organized into a world such as ours. In these formulated rules, Descartes finds the only determinate and reliable way to conceive of the laws of nature. He achieves four important effects through the procedure of grounding the conception of laws of nature in his general theory of nature in general. Whereas historians of science properly concentrate upon an internal analysis of these laws, our primary concern is with their functional relationship within the Cartesian teaching on the meaning of nature.

One fruit of this functional treatment is to prevent any metaphysical misconception about the autonomy of the laws of nature. As Descartes understands them, these laws cannot be mistaken for subsistent forces to which God and the entire course of material events must bend. Laws *of* nature are quite pregnantly laws *for* nature, in the sense of being dependent and instrumental expressions of the manner in which the divine power conveys movement to the whole field of material particles. The laws or rules themselves are intellectual formulations of the pattern followed by God in communicating motions to bodies. Hence there is engraven in the Cartesian meaning for laws of nature a total dependence upon God and an intervening act of intellectual specification, both on the part of the divine mind and on the part of the human mind developing the analytic science of mechanics. Laws of nature have no independent ontological status, therefore, but are specifying principles for determining and understanding mechanical movements in matter.

In the second place, Descartes sounds a note of epistemological self-disciplining when he regards laws of nature as general mechanical rules governing the changing relations among material particles. Such rules are not apprehended by an indolent gaze of the common-sense mind upon natural happenings, but demand arduous intellectual activity for their generalization and precise statement. Descartes expressly submits a knowledge of laws of nature to the skeptical considerations affecting every topic in his treatise on *World*, since this reflects the actual historical situation in which Montaigne and his successors included such laws within the scope of their suspensive act. Just as the basic particles of matter (the second meaning for nature) are imperceptible and gain acceptance only through a deliberate doubting of sense realism and a rational interpretation of phenomena, so also does the acceptance of laws of nature (the third meaning of nature) require a similar suspension of sense-based convictions and an interpretative use of the geometrical and mechanical modes of reason. By equating laws of nature with rules of mechanical motions, Descartes underlines the strong conceptual factor in such laws, as well as the mental discipline required to formulate and retain them even in the face of our habitual talk about animistic forces and qualities.

Yet the Cartesian notion of laws of nature is not a purely theoretical and constructural one. For the third consequence of viewing these laws in the perspective of Descartes' general theory of nature is that they share in the complexity of factors required for understanding nature. Into the meaning of laws of nature he always weaves together the factors of real causal agency, mechanical principles of explanation, and descriptive schema of intelligibility. They are closely integrated in the signification of nature-as-form, and hence they cannot be distinguished, even for purposes of analysis, as sharply as a present-day philosopher of science would prefer to do.

A Cartesian law of nature does signify some real referent or process of change in the relations among material particles. But its formal character signifies that the changes are being considered in their most general pattern, insofar as the natural processes express both the divine power at work in shaping the universe and the laws formulated in the human sciences of concrete geometry and mechanics. The correlativity between these component meanings is nicely conveyed in Descartes' remark: "I have also noticed certain laws which God has established in nature [materially considered], and of which he has imprinted notions in our souls, such that after having reflected sufficiently upon them, we could not doubt that they are exactly observed in everything which is or which occurs in the world."[11] Thus the law of inertia and other laws of movement concern real changes occurring in the world, but construed by our reflective minds as being regulated by a general procedure which God imparts to particulate matter and which we formulate into a mechanical explanation of sensible phenomena.

---

[11] *Discourse on Method*, Pt. 5 (Olscamp, p. 34). This text shows how intimately Descartes coordinates his theory of nature with his basic epistemological and methodological view of the human mind, as containing some divine sparks and primordial seeds of truth about itself and the world: *Private Thoughts* (AG, p. 4); *Rules for the Direction of the Mind*, rule 4, in Descartes, *Philosophical Essays: Discourse on Method; Meditations; Rules for the Direction of the Mind*, tr. by L. Lafleur (Indianapolis, 1964), pp. 158, 160 (cited hereafter as "Lafleur" translation).

The laws of nature not only furnish the broadest principles for an effective mechanical explanation of the world, but also contain descriptive resources. They enable us to synthesize the totality of moving particles, so regulated, into a unified description of nature. The descriptive function for the Cartesian laws of nature manifests itself in the very powerful image of nature as an encompassing machine. This holistic schema provides the interpreter of nature with a descriptive framework for harmonizing all the mechanically ordered changes and structures in the world.

This leads to a fourth and final contribution of Descartes' formal approach to nature: it saves him from being victimized by the general metaphor of nature-as-machine and the mechanistic models of particular kinds of natural phenomena. He employs the contextual image of nature as a great cosmic machine in order to eliminate occult forces and qualities from the explanation of natural events, but he does not permit this imagery to become a quasi-deity overwhelming man and blocking recognition of God. Its instrumental status, as aiding man to understand and act effectively in the world, is succinctly stated in this passage from an indirectly transcribed letter of Descartes:

> The great mechanism being nothing else than the order which God has impressed on the face of his work, which we ordinarily call *nature*, [Descartes] deemed it to be better to consider this great model and apply himself to following this example, rather than the rules and maxims established through the caprice of several armchair thinkers, whose imaginary principles produce no fruit at all because they are in accord with neither nature nor the person who seeks to be instructed.[12]

Descartes' attitude toward *la grande Mécanique de la Nature* is that it serves negatively to liberate the mind from a purely qualitative physics and a nonscientific approach to natural happenings, and positively to establish a certain proportioning between the reality of nature and the inquiring human person.

Descartes is sometimes chided for treating nature as an independent machine and then banishing our mind dualistically from its sphere of reality. But such alienation does not actually follow from the Cartesian conception of the image of nature as a great machine. Insofar as this metaphor is a way of conceiving the formal aspect of

---

[12] This *Letter of June 2, 1631*, to Villebressieu, was reported in indirect discourse in A. Baillet's classic *La Vie de Monsieur Descartes* (Alquié, vol. I, p. 294). Descartes' critics, noting his penchant for vivid imagery in his own physical explanations, accused him in turn of writing "a novel about nature." But Baillet observed that Descartes himself had already applied this phrase sardonically to his own *Monde*.

nature, it signifies a stable ordering of the movements of material particles, but not an independent entity or self-fueling engine. This covering model does not estrange human awareness from the world and does not raise up a hulking power to diminish man's active interests and values in the world. As long as the world machine is kept within the reflective context of the Cartesian theory of nature-as-form and the theory of laws of nature, it serves rather as a coadaptive bond between nature and the human investigator.

As we ponder the meaning of a machine metaphor, we gain some assurance that it incorporates a generous measure of human interpretation and imagery. Hence a cosmic mechanism can be deliberately regarded as a model for our inspection and use, as an effective sign that the reality of nature is orderly and amenable in some degree to human interpretation and use. This bonding of human thought and action to the world order, through the influence of the machine metaphor, is just what is lacking in the armchair or purely qualitative philosophers of nature. For they spin out imaginary speculations about nature, without accepting the discipline and guidance afforded by concrete geometry and mechanics. Their view of nature may seem warmly human at first glance, but it fails to show the same relevance for our human understanding and control of nature that comes from making a critical Cartesian use of the machine metaphor.

5

Fabular Man's Presence in Nature

BY developing a detailed conception of nature in conformity with the universal metaphor of mechanism, men can fulfill their active vocation in the world. For it encourages them to organize and use their practical knowledge in a manner that will make them "the masters and possessors, as it were, of nature."[13] This famous phrase in the concluding part of the *Discourse on Method* expresses Descartes' hope for attaining technical knowledge and control over the natural world, approached in the light of the image of the great cosmic machine. What must not be permitted to slip by unnoticed, however, is the important qualification attached to this ambitious program of realizing human mastery over the visible world through the resources of mechanistic science and its reflective support in the philosophy of nature. By means of these resources, men will become "as it were" (*comme*) the practical masters of the sensible universe. Through this restrictive word, Descartes reminds us that his commitment to the program is not untempered by other philosophical considerations concerning the reality of nature and the presence of man therein.

What saves him from making extravagant claims for the mechanistic explanation and its technological counterpart is nothing less than the basic Cartesian teaching on the threefold signification of nature. The formal conception of nature can never break loose from the qualifying context set by the theistic and material conceptions of nature. Hence the mechanicomorphic basis for hoping in our progressive mastery over nature cannot be absolutized without becoming like any other deification of a formal description and explanation. There is a real material structure in nature which is not wholly

[13] *Discourse on Method*, Pt. 6 (Olscamp, p. 50). Descartes is not just echoing Bacon's words about man as the servant and commander of nature, but is seeking to meet Montaigne's skeptical mockery of human aspirations toward control over nature. The Cartesian stress on continuity between a mechanical philosophy of nature and technological aspects of the new science answers this rhetorical question of Montaigne: "Is it possible to imagine anything so ridiculous as that this miserable and puny creature [man], who is not even master of himself, exposed to the attacks of all things, should call himself master and emperor of the universe, the least part of which it is not in his power to know, much less to command?" *Essays*, Bk. II, sect. 12; in *The Complete Works of Montaigne*, tr. by D. Frame (Stanford, 1957), p. 329. That a positive personalist humanism can nevertheless operate within the skeptical limits is the theme of P. P. Hallie's *The Scar of Montaigne* (Middletown, 1966).

captured in the machine metaphor and not wholly pliable to our instruments of exploitation. And insofar as nature is considered in reference to its divine source, it has a power operating in it which is not perfectly proportioned to human capacities and purposes.

To make us more reflectively aware of the exact import of the machine imagery of nature was one of the pressing reasons behind Descartes' move from his early, relatively isolated treatment of the world to his integration of the theory of nature with the rest of his philosophy. He felt the need to establish a more explicit link between the machine metaphor of nature and the epistemology of our mechanistic hypotheses and models, with their intrinsic limitations. Our possessive and manipulative attitude toward nature is qualified by some noetic conditions, which become articulated only within a fully integrated philosophy of nature. Men cannot exhaustively intuit the meaning of nature, but must strain to determine some portion of it by framing hypotheses and devising models. These latter acts of the inquiring mind do not require us to pretend that nature comes entirely within our purview or that it is ordained exclusively for our convenience. But taken in conjunction with technological advances, such intellectual acts do assure us that nature sustains our humanizing efforts of intelligence and control, even though it never functions solely as a backdrop for human desires, however Promethean.

Descartes' incomplete sections on *Man* served as more than a prolongation of the general viewpoint proposed in the *World*: they also underlined the need to assimilate his initial analysis of the meanings of nature to the ordering principles of his metaphysics and philosophy of nature. For there was an exact proportion between his new tale of nature and his treatment of man as comprising one interesting chapter within that story. Both the strength and the limitations of his general account of nature become incorporated into the application of the tale to the case of man.

As a sign of this coadaptation, Descartes was careful not to refer unconditionally to human reality but to concern himself with "the men of this new world . . . men who resemble us."[14] The fabular

---

[14] *Le Monde*, chap. 13, and *L'Homme* (Alquié, vol. I, pp. 373, 379); to be contrasted with Descartes' subsequent references to the "true man" or non-fabular human reality encountered in our actual experience (cf. Alquié, vol. I, p. 479). The probable influence of the Disneyland-like Francini gardens and grottoes (with hydraulically powered statues that moved, danced and sang) upon Descartes and the mechanistic theories of animate motion is weighed by Julian Jaynes, "The Problem of Animate Motion in the Seventeenth Century," *Journal of the History of Ideas*, vol. 31 (1970), pp. 219–234. See Descartes, *Philosophical Letters* (K, pp. 53–54).

reconstruction of nature as a whole does not leave human nature untouched, but entails a similar reconstruction of the men who are conceived as dwelling in that world. Their entire bodily structure and organic functions can be conceived in terms of a mechanical description and explanation. Using the hydraulic automaton of his day as a model, Descartes drew some fascinating analogies between the muscles and nerves of men and the familiar grottoes and fountains at Fontainebleau or Saint-Germain-en-Laye.

This mechanicomorphic view of life and the human organism helped to manifest the scope and explanatory power of the machine metaphor of nature. And as far as it went, it also held good for actually living men. Just as Descartes was ready to apply his general theory of mechanical nature to our experienced world, so was he prepared to interpret the actual human organism in terms of the mechanisms attributed to fabular man. The appearances for which the machine imagery accounts in fabular man are the same as those constituting our experience of what Descartes called "true man," that is, the actual men living in the existent order of nature.

Nevertheless, the as yet unintegrated Cartesian anthropology was not sufficiently equipped to deal with the epistemological qualifications required for the mechanistic theory of fabular man, or with its significant points of difference from a philosophical judgment about existing men. In his chapters on *Man*, therefore, Descartes took care to give several indications of the inadequacy of his separate treatment of the mechanical structures in man. He drew repeated attention to the human imagination as the operative source of all the models and guiding analogies found in his mechanistic reading of the human organism. Despite its persuasive descriptive and explanatory results, this reading could not surpass in reliability and mode of intelligibility its primary noetic basis in the imagination. Thus the entire mechanistic rendition of human life remained fundamentally vulnerable to skeptical questioning about how human perception and imagery are related to the actual being of men.

Descartes did not break off his analysis of man without pressing home the need to recast his entire theory of nature and man within a metaphysics and philosophy of nature developed to meet the skeptical challenge. He had achieved a sharp visualization of how fabular man is rendered present to nature conceived as a grand machine. But such vivid and useful visualization still left unanswered the truth question about whether the tale yields us any understanding of existential reality. Despite all the details in his diagrams, Descartes left undetermined the crucial relationship between mechanically aroused changes in the tubes and fibers of the human frame and the meaningful perceptions and judgments of actual men. We must still suppose that a living percipient man somehow embraces within his awareness *all* the components in the Cartesian sketch of sensation. This is just a special way of stating that the relationship remains obscure between Descartes' tale of nature and fabular man and his philosophical judgment about the existing natural order, inclusive of men. His early writings open a little aperture for receiving the skeptical ferret once more, and thus for seeking out a metaphysico-epistemological transformation of his entire theory of nature.

Part II
INCORPORATION OF THE THEORY
OF NATURE

Introduction

DESCARTES' ripest philosophical achievements came during the 1640's, when he laid the foundations of his metaphysics in *Meditations on First Philosophy* and made a systematic presentation of his speculative positions in *Principles of Philosophy*. In the course of this work, he transformed his general theory of the world and man from a separate physics into a *philosophy of nature*, that is, an integral phase in the total development of philosophical wisdom. Our interest does not lie in the detailed laws and explanations comprising his account of the physical world but rather in the general conception of nature itself, which was deeply affected by being incorporated into the broader principles of philosophy. The shift from a relatively unsupported tale of nature to an incorporated or philosophically contextualized theory of nature did not greatly alter the general meanings already considered. Descartes continued to think about nature in terms of its theistic, material, and formal-law bases. But in his later writings, he subjected this pattern of thinking to more careful scrutiny for its evidential grounds and the epistemological limitations surrounding its human uses.

Once Descartes' general conception of nature was brought within the scope of his metaphysical refounding of philosophy, it became subject to a fivefold modification. First, Descartes bathed it more thoroughly in the crucible of skeptical doubt, so that the need for having a metaphysical support would be recognized and effectively met. Next, he had to effect the transition from metaphysics to philosophy of nature in a manner that would secure the proper ordering of his entire course of inferences. Thirdly, Descartes felt the urgency now to spell out the limited capacity of our mind to know nature in its several aspects. And lest discouragement should set in with a recognition of such limitations, his fourth task was to show that we are able to fashion some intellectual tools for constantly improving our practical knowledge and control over nature. Finally, he sealed the epistemological circle by reflectively considering the appeal which men often make to nature as a source of their practical beliefs, including the common conviction about some finality in our world.

The task of Part Two is to attempt to determine just how deeply these five moves affect Descartes' ultimate position on nature. Even in formulating the problem in this preliminary way, however, we can see that the process of incorporation is achievable only by establishing a *many-sided* relationship between the meaning of nature and Descartes' other basic views on knowledge, metaphysical truths, and practical aims. Perhaps the most original and lasting feature of the Cartesian theory of nature lies in this interconnective relationship with so many other fundamental principles in the philosophy.

6

Methodic Doubt and the
Metaphysical Rooting of Nature

IN telling his initial tale of nature, Descartes had subjected it to a somewhat restricted form of skeptical doubt. The latter was permitted to operate just sufficiently to prevent sense perception from being the ultimate arbiter about the constitution of bodies, and thus sufficiently to make room for interpreting the bodily world in terms of mechanistic science. Yet the resulting particulate and mechanistic view of nature still remained vulnerable from two directions, showing the need for a more drastic application of doubt.

On the one side, it could be submerged in a recrudescence of naive sense realism, with its stubborn conviction that independent sense qualities and substantial forms can provide explanatory principles for understanding nature. Descartes did not make the mistake of underestimating the common-sense attractiveness of the school manuals in philosophy of nature, since they codified and gave conceptual generalization to some deep-rooted prejudices afflicting the human mind during its abusive infancy, its period of tutorship primarily to the senses. By "the senses," however, he did not mean simply the modes of organic sensory cognition but, more inclusively, those pre-reflective judgments and everyday appraisals which we make under the influence of uncritically accepted sense perceptions.[15] A very restricted use of doubt might not be powerful enough to prevent men from appealing psychologically to their common-sense beliefs, in preference to the mechanistic analysis of nature.

In another and somewhat unexpected quarter, Descartes felt that there was a real threat to his conception of nature. He realized how

---

[15] See *Replies to Objections*, nos. 2 and 6 (HR, vol. II, pp. 49–51, 154); *Principles of Philosophy*, Bk. I, sects. 1, 66, 71–76 (HR, vol. I, pp. 219, 247, 249–253). Epistemologically, the term "the senses" refers to: (a) organic states and their correlated sensations; (b) descriptive judgments analyzing the internal structure of sensory images; and (c) existential and qualitative judgments made about the reality of the sensible world, but made precipitatively and on the sole basis of such imagery. The metaphysically attentive mind must guard against making the illicit passage from (a) and (b) to (c) as a direct and sufficient route, and must regulate its existential and qualitative reality assertions by intellectual criteria for doubt-resistant evidence. Thus the disciplined mind recalls itself from the sensory and corporeal order as much as is humanly possible.

40 DESCARTES' PHILOSOPHY OF NATURE

strong an impetus toward accepting the mechanistic view came from Galileo's scientific work. But on the side of philosophical argument and interpretation—so important for winning strong assent to a new meaning of nature—Galileo left his position exposed. Despite a few philosophical sorties of his imagination, he left his theory of nature in an unintegrated condition and lacking in philosophical foundations.

Descartes itemizes briefly the three principal defects in Galileo's approach to nature, and thus tacitly alerts us to his own tasks.[16] The basic criticism is that Galileo's own method of doubt is very restricted in range. The primary role of the senses in shaping our view of nature is questioned, indeed, but only in order to give free play to a quantitative analysis of bodies in motion. What Galileo fails to realize is that current skeptical questioning extends even to his mathematical and mechanical constructions, as soon as any existential relevance is claimed for them. There is no direct handling of skeptical doubt concerning the world machine's relationship with the experienced order of nature.

In the second place, Galileo does attempt to find an indirect ontological grounding for his account of nature, by appealing to God as the divine geometer and guarantor of the objectivity of mechanistic nature. But this defense against the common-sense realists and the skeptics is ineffective, since it does not explain how we come to the philosophically certified truth about God's existence and nature. Galileo lacks the metaphysical reconstruction of our knowledge of God which Descartes deems to be essential for a philosophical defense of the meanings of nature.

[16] *Letters of November 1633, February 1634, April 16, 1634, and August 14, 1634*, to Mersenne (Alquié, vol. I, pp. 487–489, 492–501). For a comparison of Galileo and Descartes on nature and the object of physics, see T. P. McTighe, "Galileo's 'Platonism': A Reconsideration," in *Galileo: Man of Science*, ed. by Ernan McMullin (New York, 1968), especially pp. 378–380. Descartes compresses his critique of Galileo's study of nature into a sentence: "It seems to me that he lacks a good deal in that he continually makes digressions and does not stop to explain an issue thoroughly; which shows that he has not examined them [physical issues] in order, and that, without having considered the first causes of nature, he has only sought the reasons for some particular effects, and thus that he has built without foundation." *Letter of October 11, 1638*, to Mersenne (Alquié, vol. II, p. 91). On the other hand, Descartes criticizes the Roman cardinals for hazarding any pronouncement on a scientific matter, and hopes that the 1633 condemnation of Galileo will not pit the Church against the earth's movement, as it was once pitted against the antipodes. The trouble is that Galileo's critics "mix Aristotle with the Bible, and want to abuse the Church's authority in order to exert their own passions." *Letter of March 31, 1641*, to Mersenne (Alquié, vol. II, p. 323). There are criticisms of Galileo and his theological opponents in Descartes, *Philosophical Letters* (K, pp. 25–27, 98).

The third defect concerns the degree of generalization and continuity achieved in Galileo's mechanistic view of bodies. He does propose a general mechanistic analysis of planetary motions and terrestrial happenings, but its full resources for transforming our basic conception of nature are never explored. The Italian scientist remains consistently disappointing to Descartes, as soon as the general philosophical meanings of nature come up for revision. Neither thinker is prepared, however, to distinguish very sharply between the responsibilities of the scientist and those of the philosopher of nature.

That philosophical issues are very important in establishing the mechanistic conception of nature, is clearly illustrated for Descartes in the inconclusive generalization of the mechanistic outlook achieved by his friend, Mersenne.[17] The latter sought to combine a popularized version of Copernican astronomy and Galilean mechanics with his religious faith and informal sense realism, while trying to avoid fundamental epistemological and metaphysical problems. In Descartes' judgment, this evasion is ultimately unsuccessful within the modern climate of a skeptical testing of all reality claims. For it provokes the skeptic to conclude that the mechanicist conception of nature either is a purely subjective notion or else is related to real happenings only through an uncritical appeal to God and to the objectivity of one's own cognitive acts.

Such an appraisal may not greatly affect the intrinsic meanings of nature, but it does make a profound difference in how men *use* these meanings to understand and reshape their natural situation. If a mechanistic theory of nature evades rather than overcomes the skeptical challenge in epistemology, then people will always question whether the mechanical laws of movement and the machine models can yield any reliable knowledge and guidance about the reality of nature. Under these suspensive conditions, the philosophical treatment of nature can never make passage from telling a likely tale to

[17] P. H. J. Hoenen, "Descartes's Mechanicism", in *Descartes*, ed. by Willis Doney, *op. cit.*, pp. 353–368, argues that Descartes' mechanistic world system rests primarily on a metaphysical denial of intrinsic change in bodies, from which follow his sole acceptance of local movement and his denial of activity to extended body. See R. Lenoble, *Mersenne, ou la naissance du mécanisme* (Paris, 1943), pp. 336–449, on how the Descartes-Mersenne relationship was strained precisely because of the former's insistence on a metaphysical basis for mechanism. R. H. Popkin, *The History of Scepticism from Erasmus to Descartes* (New York, 1968), pp. 142–143, points out that Mersenne himself recognized the new physical science as developing a distinctive sort of cognition which is apparential and probable, useful and not dependent on metaphysical certitudes. Descartes' own methodological adaptations to such cognition are considered below.

furnishing at least some certitudinal understanding of the real structure of nature and some definite encouragement for human control over nature.

Given the interrelated pressures exerted by sense realism, Galileo, and also Mersenne, then Descartes felt obliged to meet the skeptical challenge head-on, by supplying the theory of nature with a metaphysical foundation and order of reasoning. That is why he did not permit the doubting process, which fills the opening pages of his *Meditations on First Philosophy*, to stop short with questioning the everyday view of the world and its categorization in the school manuals of natural philosophy. Instead, he extended its range deliberately to include the seventeenth-century scientific conception of nature, together with its intellectual bases in mathematics, mechanics, and creative imagination. The methodical inclusion of these cognitive tools within the movement of doubt meant that the Cartesian theory of nature, worked out in the *World* with their aid, was now itself being submitted to the radical skeptical test for its existential relevance.

In the *Meditations*, Descartes reaffirms that his own general meanings for nature have to be subjected to doubt, in view of the classical skeptical queries about *which* nature is being discussed and *whether* nature can be known at all to exist. But by the same token, he now gains some specific assurance that his metaphysical refounding of knowledge will also benefit his theory of nature. The latter is now able to receive some principles of being and knowing upon which its existential bearing hinges, to place the necessary epistemological restrictions upon its several components, and thus to transform itself into a philosophically integrated interpretation of nature. The very proportion between Descartes' universal doubt and his universal reconstruction of the principles of human knowledge secures a systematic place for the philosophy of nature within his synthesis.

Cartesian scholars have debated about the exact relationship between his metaphysics and his philosophy of nature. Our study of the gradual formation of the meanings of nature and their eventual submission to methodical doubt suggests that both these disciplines must subserve the comprehensive requirements of Descartes' movement of philosophical reflection, and hence that their comparison must always be mediated by the idea of a systematic search for wisdom. Within this perspective, metaphysics enjoys a *foundational* primacy. For its task is to render secure the evidential basis for those guiding noetic principles (whether concerned with existing realities, with the criterion of truth, or with general truths of relationship) which are involved in every stage of philosophical reasoning. Yet the manner of their involvement is not univocally the same for every portion of philosophical wisdom, but depends as much upon the needs of the inquiry in each specific area as upon the metaphysical principles themselves.

Thus Descartes does not have to generate his basic meanings of nature through some sheerly abstract *deduction from* metaphysics. But he must show that his conception of nature does indeed *conform with* the general conditions of evidence, reasoning, and existential judgment required in a humanly elaborated philosophy. Metaphysics performs a foundational function for the theory of nature, precisely insofar as it subjects that theory to methodic doubt, determines its orderly connection with the primal truth of the Cogito or existing thinking self, and thus assimilates the meanings of nature to the growing body of philosophical certitudes. It is in this definite but restricted sense that Descartes declares that "these six *Meditations* [*on First Philosophy*] contain all the foundations of my physics," and that "I could not have found the foundations of physics, if I had not sought them by this way" of linking the meanings of nature with the metaphysical truths about the Cogito and God.[18]

It is well to avoid two extreme interpretations of these remarks. One view is that they represent the standpoint of *a priori meta-physicism*, which relates the theory of nature so passively to metaphysical principles that it is no more than their deductive prolongation into a study of the world. This reading of Descartes suffers from a threefold drawback.

First, it forgets that Descartes develops the meanings of nature directly through a mathematical and mechanical analysis of our physical experience. The function of metaphysics is not to render this distinct analysis superfluous or to supplant it with other meanings. Instead, metaphysics broadens the entire context of the inquiry into nature, so as to face the existential question and relate the complex signification of nature to the other reflective interpretants of human experience.

Next, the interpretation of metaphysicism rests upon a very narrow account of Cartesian deduction. The latter is not limited to a formal entailment from axioms and definitions, but enlists various resources of the human mind in exploring and interrelating the specific domains

18 *Letters of April 15, 1630, and January 28, 1641*, to Mersenne (Alquié, vol. I, p. 259; vol. II, p. 316); cf. *Entretien avec Burman*, ed. by C. Adam, *op. cit.*, p. 74. On the physicism-metaphysicism issue among older Cartesian scholars, see A. B. Gibson, *The Philosophy of Descartes* (London, 1932), pp. 47–56. The basic statements are in Descartes, *Philosophical Letters* (K, pp. 10–12, 93–94).

in nature. In Chapters 8 and 9 below, we will find Descartes explicitly requiring several kinds of reasoning and modes of certitude in the study of nature. These differentiations must be incorporated into any "deductive" relationship holding between his metaphysics and philosophy of nature.

And thirdly, metaphysicism fails to specify the sense in which metaphysics "gives foundations" to the theory of nature. Admittedly, we find it strange today to think that a scientifically developed account of nature needs a metaphysical context or any other sort of intellectual support. Everything depends, however, upon how far we push the questions about the enveloping human presuppositions of the theme of nature. Descartes' talk about metaphysical foundations registers his sensitivity toward such probings at the metascientific level. Metaphysics furnishes foundations precisely insofar as it methodically exposes the Cartesian meanings of nature to skeptical query, and then shows the relevance of its own findings to the existential aspects of nature.

At the other extreme is the position of *physicism*, which holds that Descartes makes a merely instrumental use of metaphysics to serve the ends of his philosophical physics. But it is just as self-defeating to take a purely tool approach to Cartesian metaphysics as to take a purely derivational view of Cartesian physics. Coadaptation and functional service are present on both sides, not in order to reduce either one of these disciplines to the status of a pure tool of the other, but in order to bring out their maximum relevancy to each other, in pursuing the common aim of philosophical wisdom. In point of fact, the metaphysics of Descartes cannot be fully understood on the premiss of its being no more than the servant of his philosophy of nature. For his metaphysics possesses a distinctive structure of problems, evidences, and order of truths pertinent to its own domain. These metaphysical findings do indeed have some implications bearing upon the inquiry into nature, but there are other consequences which cannot be developed solely in respect to the regional requirements of the theory of nature, but which must take account of the broader tasks of the philosophical search for wisdom.

Actually, Descartes is motivated neither by metaphysicism nor by physicism in relating his conception of nature to his metaphysics. Rather, this relationship is internally demanded by the reciprocal proportion between the existential aspects of his meaning of nature and the universal scope of his methodic doubt, in respect to all existential assertions. Such doubt cannot be met simply by appealing to the trustworthiness and fruitfulness of the mathematical and mechanical concepts involved, since the question agitated by the skeptics concerns precisely whether there is existential significance in such concepts, even when they are shown to be theoretically coherent and practically helpful.

An interview which Descartes accorded to a young admirer and reporter, Francis Burman, emphasizes the point that a genuinely philosophical approach to nature must determine its existential connotation.

> The entire and universal object of mathematics, and whatever it considers therein, is true and real being [*ens*], and has a true and real nature, no less than does the object of physics itself. But their difference consists solely in this that physics considers its object not only as true and real being but as actual and existing as such [*actu et qua tale existens*]. But mathematics considers its object solely as possible, and one which does not actually exist in space, but yet can exist.[19]

Whereas the Aristotelian manual writers studied this difference between mathematics and philosophy of nature solely in terms of the degrees of abstraction, Descartes employs it to emphasize the vulnerability of the theory of nature to skeptical doubt and hence its corresponding need for a metaphysical foundation. In an adequately conceived philosophy of nature, the relevant mathematical and mechanical principles become fully physicized through their adaptive reference to that which actually exists in the spatial world. And since Descartes gains indubitable knowledge of the existing extended world only through a metaphysical study of the human self and God, he regards such existential knowledge as indispensable for philosophy of nature. Only through this epistemological reflection can the passage be surely made from nature as an essential meaning (or *ens*) to nature as an actually existent reality, thus responding to the skeptical question of whether any aspect of this meaning concerns that which we know to be *actu et qua tale existens*.

In order to press home more vividly the need for his epistemological context, Descartes also considers the case of the atheistic mathematician.[20] Just as the figure of the conscientious atheist was used during the second half of the seventeenth century to explore problems in moral and political philosophy, so was it used during the first part of that century to clarify such speculative issues as the existential bearing of one's view of nature. Descartes does not dispute

[19] *Entretien avec Burman*, ed. by C. Adam, *op. cit.*, p. 54.
[20] *Replies to Objections*, Nos. 2, 3, and 6 (HR, vol. II, pp. 39, 78, 245). Cf. *Meditations Concerning First Philosophy*, Bk. V (Lafleur, pp. 124–125), and *Principles of Philosophy*, Bk. I, sect. 13 (HR, vol. I, p. 224).

the professional competence of the atheistic mathematician, but inquires whether his lack of assent to God enables him to withstand the doubting process enough to preserve the mathematical interpretation of nature from skeptical attack.

It is granted that the preliminary form of methodic doubt will leave the atheistic mathematician untouched, since he does not concede the existence of a God who may be deceiving him. But his armor will be pierced as soon as the doubt assumes its most radical form in the hypothesis of the evil spirit. To overcome the postulate of a malign and very powerful intelligence, a *spiritus malignus* aggressively bent upon distorting the mathematician's present power of judgment and warping his memory of past evidence and reasoning, the full resources of metaphysical reflection must be summoned.

In using the symbol of the atheistic mathematician, Descartes' intent is not emotive but epistemological. He does not seek to arouse abhorrence, but simply a recognition that every human modality of knowing, including the mathematical, is exposed to some form of doubt. The exposure increases in the measure that the knowing becomes involved in existential affirmations, as it inevitably does in a philosophically significant treatment of nature. In the same degree that mathematics gets concretized and physicalized in the mechanistic view of nature, it also becomes existentialized and brought within the range of the epistemological questioning symbolized by the evil-spirit hypothesis. Hence the reflectively established truths about the human Cogito and the all-powerful, veracious God are both needed, if the ideal constructs and the use of memory and imagination in mathematics and mechanics are to yield any unassailable truths concerning the existing order of nature.

What more specific benefits can the philosophy of nature expect to reap from its alliance with these metaphysical truths? Descartes' basic response is that several aspects of the study of nature, previously left in a suppositional condition, can now be rendered evidential. The transition from supposition to founded truth is most clearly seen, wherever the doctrine on God is implicated either indirectly or directly in the Cartesian conception of nature.

There is an *indirect* involvement of God in the philosophically tested conviction that an extended material world can exist and does in fact exist.[21] When the Cartesian philosopher of nature is questioned on this point, he must seek support not only in the internal coherence of the idea of a mechanical world, but also in the relationship which

[21] *Principles of Philosophy*, Bk. I, sects. 69–70, and Bk. II, sect. 1 (HR, vol. I, pp. 248–249, 254–255); *Meditations*, Bk. VI (Lafleur, pp. 126–134).

such a world bears to the meaning and reality of God established in metaphysics. For assurance about the existential possibility of this world, he has to consider both the inner consistency of the mechanistic world view and the relationship of feasible production holding between the creative power of God and an order of mechanically moving particles.

As for the further judgment that nature as a mechanical order does in fact actually exist, Descartes' original skeptical analysis of sensation forbids him to ground it ultimately in sense experience as a direct truth source. Instead, he justifies it indirectly in function of the theory of natural belief, whose complexities will be examined in the last chapter of this study. The sole point to notice now is that inclusion of our conviction in an existing order of nature among our natural beliefs leads to a metaphysical involvement of the divine veracity. The massive and ineradicable quality of our natural belief in the existing material world implicates the veracity of God. A benevolent and truth-dealing God, who is metaphysically established to be the maker and sustainer of our human reality in its permanent and inescapable traits, would not deceive us irremediably on this vital conviction. Thus the metaphysical inquiry into the infinite benevolence and veracity of God is relevant, in the long run and yet in quite specific fashion, for the existential aspects of Descartes' theory of nature.

There are some other facets in this theory which are more *directly* related with his metaphysical approach to God. This linkage is most noticeable in the Cartesian conception of laws of nature, which we have seen to be so closely associated with the problem of the more and the less eternal truths. Some features of this view of laws of nature require a correlation, not just with a general affirmation of the divine existence, but with a metaphysically specified inference concerning the divine attributes.

It is only because of the supporting context of the theory of divine attributes that Descartes is metaphysically certain about the ability of a few core points, of high generality in his notion of laws of nature, to withstand skeptical assault. These doubt-resistant points concern the mode of intelligibility of laws of nature, their enactment status and causal significance, and their immutable and universal character.

(a) By taking the most general laws of nature to be rules of mechanical movement formulated by the human sciences of geometry and mechanics, Descartes seeks to show not only that natural reality is accessible to human intelligence in some degree but also that its laws constitute a thoroughly *dependent intelligibility*. The dependence is based immediately upon the human sciences involved in the formulation of laws of motion for material particles, and is ultimately founded upon the divine intelligence and will. This ultimate or metaphysical dependence of the "less eternal" truths upon God's mind and will serves a dual purpose in Cartesian thought. It operates both as a limit against the naturalistic tendency to absolutize the laws of nature, and also as an assurance that well-formulated general rules of mechanical movement do bear upon the world which is actually produced through divine creation. Thus the dependent intelligibility of laws of nature upon the creative God encourages us to use them for exploring the world in which we actually live and work.

(b) In consequence of Descartes' refusal to permit a distinction in God between intellect and creative will, the reference of laws of nature to the world of moving bodies always has an active aspect, and not just a cognitive one. His talk about how God impresses the laws of nature (nature-as-form) upon the material subject (nature-as-matter) is not merely a legalistic analogy drawn from the role of laws in human society. Basically, he views the laws of nature as being *enactments* for the material world, because of his metaphysical position that the divine intelligence and the creative-conservative power of the divine will are indistinguishably one. The meanings enshrined in the most general laws of movement are enacted and imparted meanings, ones that are causally effective in the changes and dispositions of material particles. Even when he must stress the descriptive and hypothetical condition of so much of our particular investigation of nature, Descartes never totally eliminates the enactment and causally efficacious character of the most general laws of nature. What prevents this background from ever slipping out of sight is its anchorage in the metaphysics of the divine attributes.

(c) This same context accounts for the peculiar shape in which the problem of universal and immutable laws of nature presents itself in Cartesian thought. There is unusual sensitivity about the objection of theological and scientific critics that, if the divine intellect and will are so closely one and if the latter is free, then a consequence seems to be that God may place regional restrictions upon the laws of nature and may indeed change them quite arbitrarily. Descartes will not permit such reasoning to weaken either his metaphysics itself or his confidence in the value of a theistic metaphysical context for his philosophy of nature. He traces the objection back to a separatist consideration of the divine will and freedom by themselves alone, whereas in his account of the creative act these principles must always be considered in the concrete oneness of all attributions in the active being of God. His will and freedom cannot be separated from His infinite and immutable power of essence.

In freely establishing formal laws of nature, God also specifies *how* they will regulate the world: universally and immutably. The act whereby God determines the broadest laws of nature and communicates them to matter is characterized as free, infinitely powerful, and immutable in sustaining what has been freely determined for regulating mechanical movements. By reflecting upon the unlimited power of God in this foundational enactment, we have some assurance that the laws established for our world do have a universal range and do penetratingly regulate every movement of material particles. Similarly, we gain metaphysical support for the persistence of the laws of movement and rest when we view such persistence as being a participated expression of the unchanging ordination of God. Neither the universality nor the immutability of the laws of nature ceases to be participative and dependent in respect to the creating-conserving God. Hence divine freedom and stable laws of nature do not cancel each other out, but rather are related as creative source and effective expression.

Descartes is well aware that problems of this sort need not bother Galileo and the practicing mathematical physicists. Even his own statements of the laws governing conservation and transference of movement are not initially formulated and derived by an abstract deduction from the theory of God's perfections. Being rules for understanding nature in its formal aspect, they must be fundamentally proportionate to the basic mechanical concepts and the particulate, corpuscular view of matter. Nevertheless, the philosopher of nature is one who *does* expose himself deliberately to skeptical questioning about the reality reference and scope of mechanical laws of nature, as well as their compatibility with working theistic convictions. Hence at least Descartes himself, meeting these issues, has to become concerned with showing the continuity between his metaphysical view of God and his mechanical view of nature. Their relationship is not otiose from the standpoint of a philosopher of nature who seeks real relevance, as well as logical consistency and comprehensiveness, for his most general laws of nature.

In the following passage, Descartes highlights this mutual bond between the metaphysics of divine attributes and the primary rules of conservation and transference of mechanical movement in nature.

I showed what the laws of nature were. And without leaning my reasons upon any other principles but the infinite perfections of God, I tried to demonstrate all those laws about which we might doubt, and to show that they are such that even if God had created many worlds, there would have been none of them where these laws failed to be observed. . . . For what more firm and more solid foundation could one find to establish a truth, even if one resolved to choose it to one's own liking, than to take the very firmness-and-immutability which is in God? The fact is that these two rules [conservation of the same quantity and state of movement, and transference of movement] follow manifestly from that [truth] alone that God is immutable, and that acting always in the same way He always produces the same effect.[22]

Under conditions of radical skeptical doubt, we can still maintain the real reference and the universal range and persistence of the mechanical laws of nature by considering them in their harmonious relationship with the metaphysically established conception of the infinitive, creative God.

The divine perfections serve here as an ultimate resolutive principle for our thinking about nature in terms of unchanging and universally valid mechanical laws. Such a principle does not dictate the internal content of laws of nature, and does not constitute the only standard to which these laws must be rendered proportionate for descriptive and explanatory purposes. But it does furnish contextual assurance (and, in this sense, a demonstrative foundation) for the mechanistic conception of nature, within the actual historical condition of its being submitted by the Cartesian philosopher of nature to skeptical fire.

Hence for Descartes there is no conflict in principle between his theism and his mechanistic interpretation of nature, but rather a mutual proportionment which does not exhaust the significance of either position. On the other side, the Cartesian mechanization of nature involves a definite theization of the grounds for maintaining the reality and persistence of the first laws of movement. Our assurance is strengthened in this respect by regarding these laws in continuity with our metaphysical reflection on the self and the creative freedom of God. But conversely, the idea of nature helps to

---

[22] *Discourse on Method*, Pt. 5 (Olscamp, pp. 35–36); *Le Monde*, chap. 7 (Alquié, vol. I, p. 357). Descartes follows different sequences in deriving the three basic laws of the conservation, transference, and direction of movement, in the two classical texts: *Le Monde*, chap. 7 (Alquié, vol. I, pp. 351–359), and *Principles of Philosophy*, Bk. II, sects. 37–42 (AG, pp. 216–219). A. Koyré, *Newtonian Studies* (Cambridge, 1965), p. 70, remarks that the Cartesian state of movement refers back to divine conservation, whereas the Newtonian state refers to a power resident in bodies as such. Consult R. J. Blackwell, "Descartes' Laws of Motion," *Isis*, vol. 57 (1966), pp. 220–234, on how Descartes must refer conservation of movement to divine immutability rather than to material nature, because he lacks the Newtonian concept of a body's mass.

open out the structure of Cartesian metaphysics itself. If it were not for the two-way corridor of implications developed between the theory of God and that of nature, then perhaps Descartes' quest for philosophical wisdom would not have given such prominence to the investigation of the existing, moving, material world of sense experience. To St. Augustine's declaration: "I desire to know God and the soul," the French philosopher always adds: "and nature as well." The demands of the theme of nature prevent Descartes from becoming lost in a bottomless well of human subjectivity. His metaphysical study of the thinking self and God does not seal itself off, but becomes foundational for an exploration of nature and its mechanical laws as a major task in philosophy.

7

A Crucial Transition

FOR all of his metaphysical critique of an independently based philosophy of nature, Descartes found it advisable to maintain constant vigilance over his reordering of the philosophical disciplines. Since his incorporative approach to nature ran against the tide of accepted thinking, it carried an inevitably untimely and reforming aspect which invited challenge at every stage of his theorizing. The danger of slipping away from moorings in the Cartesian theory of being and knowing was perhaps greatest at that point in the *Principles of Philosophy* where a systematic transition was being made from Part I to Parts II–IV, that is, from the metaphysical principles of human knowing to the theory of material things in general and in specific forms. Here, if anywhere, the linkage between metaphysics and philosophy of nature might be declared too contingent and artificial to keep the latter firmly within the order of inquiry proposed by Descartes. Hence it was here that he himself devoted so much careful argumentation aimed at safeguarding a relationship which he deemed essential for the philosophical development of the theme of nature.

To appreciate his philosophical craft in treating this issue, we must look closely at the initial three principles in Part II where the general discussion of material things is broached. More important than the particular matters dealt with under these headings is their conjoint purpose of determining the general context and limitations within which every subsequent facet of the philosophical theory of material nature must operate. Since together they determine the general setting for that theory, we must consider them in synoptic fashion in order to catch Descartes' own grand strategy in placing them at the outset of his account of material things.

The three liminal principles of Part II of the *Principles* are formulated in this careful fashion:

[1] By what reasons the existence of material things is known as certain. . . .
[2] By what reasons the human body is also known to be closely conjoined to the mind. . . .
[3] That the perceptions of the senses do not teach us what is

really in things, but what aids or impedes the human composite.[23]

Descartes hopes that the conscientious reader will come to apperceive these statements in their dramatic unity. For standing at the turning point of his philosophical inquiry, as it moves from a study of oneself and God to that of material nature, their work is to suggest some good reasons for hewing to this order of inquiry and hence for refusing to sever the theory of nature from its metaphysical footing.

There are at least five functions which these three topics together fulfill in relation to the Cartesian philosophy of nature. Their first task is to *anthropologize* the entire theory of material things, insofar as it becomes assimilated to the continuous course of Cartesian philosophical inference. This does not mean that the integrity of the theory is to be destroyed by making it project man's fancies or cater to his whims, since it is against precisely such intellectual corruption of our thinking about material nature that Descartes aims his polemic against substantial forms and extrinsic finalism. Rather, the anthropologizing of the meaning of nature signifies that a noticeable reorientation occurs when we pass from a scientific treatment of bodies and a philosophical analysis conducted apart from epistemological questions to the Cartesian metaphysical setting. The latter secures its mark upon every phase in the study of bodily nature by relating it explicitly to the human self, the claims of knowledge, and the peculiar reality of man in the material world. That such an all-pervading reference of the theory of material nature to the problem of human selfhood within nature can be achieved, without lapsing into fantasy and animism about material things, is the challenge which Descartes accepts and the difference which it makes to develop an incorporative philosophy of nature.

The disciplined quality of this anthropological reference can be recognized already in the three threshold principles. Principle [1] directs our attention at once to the question of existence. In a scientifically developed theory of bodies and motions, this question would be relatively insignificant or perhaps would never be raised. But it becomes so central in Descartes' philosophical treatment of bodies that it belongs quite appropriately among the shaping general issues which cannot be delayed. Similarly, principle [2] has an odd look about it from the standpoint of a purely physical approach to

[23] *Principles of Philosophy*, Bk. II, sects. 1–3 (HR, vol. I, pp. 254–255). Since an improved English translation of this work is long overdue, it is specially advisable in the present section to consult the Latin text of *Principia Philosophiae*, in vol. VIII–1 of *Oeuvres de Descartes*, ed. by C. Adam and P. Tannery (Paris, 1964), pp. 40–42.

material things, and yet is quite properly placed in the Cartesian philosophical ordering of problems. Within the former framework, a distinctive question about the human body would not be a matter of principal concern at all, but would fit into a definitely subsequent and modest niche, awaiting fuller development in a special treatise on life and man. But Descartes alerts us, at the very outset of his philosophy of the material world, to the peculiar importance he attaches to the human composite and the intimacy of the mind-body union. The philosopher of nature's orientation and priorities are already being announced at this early stage of the inquiry. And lest this fundamental correlation between the theory of material things and that of the human organism be slighted, principle [3] provides some reinforcement of the philosophical reshaping of physical theory. Not only does it insert a word of epistemological caution against any naive identification of theoretical structures with world realities, but it also gives an otherwise unusual prominence to human pragmatic considerations in the study of bodily motions and qualities.

As a second cooperative service, these three principles *modalize* the meaning of "to know," and do so precisely at that stage in the whole inquiry where such modalization must be clearly recognized. Different considerations must be advanced in support of the claims to know that material things exist, that the human body is closely united to the mind, and that sense perception is instrumental to man's needs. In his methodological treatises and his program statements about reaching human wisdom, Descartes stresses the use of the *same* powers of mind and the attainment of unified understanding. But there are also the basic *differentiations* introduced by the relative ordering of a topic to the truth of the Cogito, by the complexity of the matter under investigation, and by the varying relations between the human composite and the other regions of reality.

It is specially incumbent upon Descartes to underline these internal qualifications upon the claim of knowledge when he is making the delicate passage from metaphysics to the rest of his philosophy. These three propositions help to remove any illusory expectation that philosophy of nature and the practical areas will yield a knowledge entirely homogeneous with that of metaphysics. They prepare us for the use of many methodic tools not required in the metaphysical principles and hence for a pluralization of certitudinal assents, not all of which can be in the same mode as those in metaphysics and the basic physical principles. Thus although Descartes does not rely as heavily as the schoolmen upon a doctrine of analogy in knowledge, he sees the need for internal differentiations and hence uses the occasion of this opening of his account of material things to dispose the reader for expecting such cognitive adjustments increasingly in the development of his philosophical view of the material world.

Just as our three related principles have this prospective aspect, so also do they have a *retrospective* aspect which constitutes their third main task. For when we ask about the founding evidence concerning the existence of material things, the nature of the mind-body union, and the cognitive range of sense perception, the import of Descartes' systematic development of these topics is that some important sources of the evidence lie outside of the scope of the theory of nature itself. These are "look elsewhere" questions, insofar as they point back to some metaphysical argumentation for at least part of the premises required for their own settlement. What this means on the broader issue of philosophical order, however, is that no absolute beginning can be made with the philosophy of nature. There must be some theoretical carryover from the metaphysical principles of knowing and being into the discussion of the principles of material things, so that the relationship between these two sets of philosophical principles is not entirely optional but involves some doctrinal precedence of the former over the latter.

Descartes will not permit this point to be made only implicitly or in passing. In his clarification of principle [1], he notes how a defense of our existential conviction about the material world depends upon establishing the veridical nature of a clear and distinct perception of the idea of matter, with its objective reference to a real existent having extension, figure, and motion. This in turn depends upon our seeing that "it is clearly repugnant to the nature of God that He be a deceiver, as has been previously explained. And hence here we must unreservedly conclude that there exists a certain extended thing."[24] Again, in discussing the union of mind and the human body in principle [2], Descartes merely sketches the elements of his position and closes with this pointed remark: "But a more accurate explanation of that thing does not belong in this place." Finally, he qualifies his elaboration of principle [3] by stating that his account of the pragmatic character of sense perception is "sufficient" to develop a critical attitude and to rely "here" (in the theory of bodies) upon intellectually tested notions. All this positional language is intended to disclose the relationships leading back to the foundations of

---

[24] The statements and words quoted in this paragraph are from the body of the same three principles cited in n. 23.

metaphysics, as well as forward to more specific treatments within the philosophy of nature.

The fourth common feature of the principles under comment is that they raise questions of the *how-do-we-know-for-certain* kind. Once more, the directing of our attention to the certitudinal quality of our assent is an indication from Descartes that more is at stake than a restricted physical theory, and that more is being demanded than an epistemologically and metaphysically ungrounded philosophy of nature can supply. To appreciate the distinctive sense in which he reopens the question of certitude as his inquiry moves into material nature, we must examine the opening statement under principle [1].

> Although no one does not persuade himself sufficiently that material things exist, since however this has been called into doubt by us a little before and is numbered among the pre-judgments of our first age, it is now our task to investigate the reasons through which it is known as certain.[25]

This is a dispositional statement in respect to the entirety of Parts II–IV of the *Principles*, not just in respect to the existential issue at hand. For it conveys Descartes' general plan of raising the certitude question throughout our course of philosophizing about material nature. He finds it just as necessary to unsettle our ordinary convictions in this sphere as in that concerning the human self and God, so that some degree of matching rigor can be introduced into the philosophy of nature.

For that purpose, he distinguishes carefully here between two human conditions of certitude and sweeps our theorizing about nature into his ongoing process of passing from the one meaning of certitude to the other:

(a) Men begin uncritically with certitude in the sense of *self-generated persuasion*. This is a conviction grounded solely in the psychological and cultural conditioning of the human understanding to believe in some massive and overwhelming realities, such as the existence of the material world. Persuasional certitude arises out of our individual and social pre-judgments and never submits itself to the testing process of radical doubt. Beliefs held in this way can become highly organized, however, as in the case of the philosophies of nature found in the school manuals. These philosophies of nature are the outcome of a codification and canonization of persuasional certitudes about material nature. Hence they cannot be criticized effectively in their own systematic terms, but only by being traced

---

[25] *Principles of Philosophy*, Bk. II, sect. 1 (HR, vol. I, p. 254; Adam-Tannery, p. 40).

back to their epistemological roots, where the contrast with another meaning of certitude can be grasped.

(b) For Descartes, the philosophical life begins when we start passing from self-generated persuasion to *evidential assent*. Certitude in the sense of evidential assent is not a mere prolongation or variation upon the persuasional kind, since we arrive at it only in the degree that we use methodic doubt to criticize our already operative convictions. The methodically ordered, guarded, and modalized assent which survives the testing for grounds of evidence is certitudinal primarily by reference to those evidential grounds, and only secondarily by reference to the mind's act of firm adherence. The psychological aspect of certitude is not eliminated, but it is made to justify itself as a subordinate element in an assent act regulated by doubt and proportioned to evidential perception.

Just as Descartes requires this reordering of firmness to evidential perception in the case of the philosophical theory of the human self and God, so does he want to induce a similar transformation in the philosophy of nature. He does not lust after certitudes at all costs. On the contrary, he encourages the sometimes painful process of disengaging ourselves from uncritically held psychic and social persuasions and of restricting our certitudinal claims in philosophy of nature, as well as in metaphysics, to those assents warranted by our evidential perceiving. After 1650, unfortunately, both the Cartesians and their skeptical critics tended to wash out the philosophical discriminations underlying this stand on certitude.

The fifth and perhaps overriding aim of the three principles being examined is to keep the transition from metaphysics to philosophy of nature oriented toward Descartes' own *ideal of the unity* of philosophy. As a linguistic and conceptual cue to this relationship he employs two words that are unusual in a philosophical setting: "first age" or "childhood" (*prima aetas*), and "mature reason" (*matura ratio*). The former term appears in the text quoted just above, whereas the latter serves as the final word of Part I of the *Principles*. There, Descartes observes that "in those matters about which divine faith teaches us nothing, it would be least fitting for a man-as-philosopher to accept anything as true which he has never ascertained [*perspexit*] to be true, and to trust more to the senses, that is, to judgments made without consideration in his infancy, than to mature reason."[26] When this sentence is linked with the three introductory principles of Part II, it enables us to appreciate their systematic import for Cartesian philosophy as a whole.

[26] *Ibid.*, Bk. I, sect. 76 (HR, vol. I, p. 253; Adam-Tannery, p. 39).

E

After having reached a peak of maturely considered philosophical judgments concerning the human self and God, Descartes seeks to avoid slipping back to a pre-critical level on the theme of material nature. What scandalizes him is not the fact of the composite union of mind and body in man, and not even the fact of organic growth as affecting man's exercise of intelligence. Rather, his criticism falls upon abusive infancy, upon an age of childhood which is unduly prolonged for the individual and the human race. (Kant regards Enlightenment as man's release from such self-incurred tutelage.) Such a stunted condition of mind obtains for those men and those philosophical schools which regulate their conceptions of reality by "the senses." We have seen that this term is to be taken in the highly qualified Cartesian sense of judgments formed without methodic self-criticism and without consideration of the prime rule of assenting, in philosophy, only to that which one personally perceives and ascertains. The infantile refusal to reinspect one's primordial beliefs is specially difficult to dislodge in questions on material nature, so that here the struggle toward philosophical maturity can be lost in the confusion between venerable prejudices and evidence-supported assents.

But Descartes is not despairing at this juncture of his work, and the basis of his hope is conveyed in the two odd terms: "first age" and "mature reason." In our very recognition of naively realist and sensist concepts of nature as being instances of arrested reflection, we can assign them to the childhood beliefs of the human race and can look forward to a growth beyond the first age of man in nature. The promise of further growth is backed up, more immediately, by the example of Cartesian metaphysical criticism of uncriticized beliefs about the self and God, which can now be extended into the domain of beliefs about nature. But our deeper confidence must rest upon an interpretation of human reason itself as being involved in a living process of growth and reflective maturation.

It is the realization of our reason-as-maturing which ultimately cannot be thwarted by a refusal to reconsider early beliefs concerning nature. The *bona mens*, or the living thrust of mind toward wisdom and happiness, urges men to reconstitute their concepts of nature within the sustained course of Cartesian reflection and its metaphysical principles. Engagement in the maturation process for human reason is the powerful ideal upon which Descartes relies for inducing men to exercise the philosophical rule of disciplining their assent concerning material nature by their own reflective perceptions and ascertainments

One dividend which we gain from this commentary concerns the status of the famous metaphor of the tree of wisdom. Descartes is by no means unaware that it is a metaphor and hence that, taken by itself, it is exposed to a basic objection. Being a conjunctive and hence essentially complex image, a metaphor has component parts which can be subjected to our imaginative variation. In this instance, it might be argued that the imaginal components representing the different parts of philosophy can be varied at will, and especially that the order between metaphysics and philosophy of nature can be reversed so that the former is the central trunk and the latter is the roots of the tree of philosophical wisdom. Descartes might concede this criticism, but only under the doubtful supposition that the metaphor can indeed be taken by itself and regarded as having a self-founding validity.

What he could not concede—and the analysis of his intention in the three principles just studied emphasizes the point—is that the metaphor does, in fact, enjoy autonomy and afford an independent basis for determining the proper order between metaphysics and philosophy of nature. Instead, Descartes is offering an *outcome-image*, a vivid way of recapitulating and expressing in easily apprehensible form his pathway of argumentation on the ordering of the philosophical disciplines. The tree metaphor is a concrete reforming instrument for focusing his complex analyses of the issue, including the fivefold considerations involved in the transition from the first to the remaining parts of the *Principles*. Given this argumentation in favor of one specific ordering and against other kinds, there is a definite heuristic principle for understanding that and why metaphysics must hold the root position, and philosophy of nature that of the mediating trunk, in the tree of philosophical wisdom. This recapitulative metaphor does not leave the question of philosophical order open to purely optional rearrangement or to associations based solely upon the organic model used. In order to underline the critically regulated outcome-character of the wisdom tree image, Descartes offers it in the Letter to the Translator of the French edition of *Principles of Philosophy*, rather than in the original statement and argued body of that work.

8

Tempering the Claim to Know Nature

WHAT sets off Descartes himself from most Cartesians and from the standard picture of the rationalist thinker is his consistent refusal to convert the foundational use of the metaphysics of self and God into a basis for claiming to have exhaustive insight into the essential workings and different species in nature. Metaphysical considerations help to disarm skeptical doubts concerning the scope and ontic reference of the mechanistic conception of nature in general. But they do not underwrite the extreme claim that this general view of nature satisfies our effort to understand the different kinds of reality in the natural world.

At this point, we must catch the epistemological significance of Descartes' prime distinction between the *general* meanings of nature and the *progressively more determinate* meanings oriented toward our experienced world. Even after we are able to develop a complex general significance for nature in terms of God, material particles, and the laws of movement, we are not in possession of a fully satisfactory concept of nature. For this complex general sense of nature is not terminal, but instrumental, in our continuing study of experienced natural reality. The concept of nature must be rendered constantly more determinate and concrete. This requires us to seek an ever closer continuity between the general meaning and the more particularized study of the appearing world, the human environment of our earth, and the kinds of bodies we encounter and use.

By acknowledging that this transition is needed for the intrinsic perfecting of the meaning of nature, Descartes avoids a facile mechanistic rationalism and tempers his philosophical claim to know nature. The passage cannot be made simply by reading off the more specific and determinate traits of nature from the complex general meaning, regarded as a sufficient axiomatic premiss. Since there is genuine scientific discovery and enrichment of the meaning of nature, as it becomes specified to include the different phenomena and particular structures in our world, the intellectual tools and conditions of such scientific inquiry must become incorporated into the conception of nature and must impose certain limits upon it. Thus Cartesian reflection upon scientific methods is pertinent for the theory of nature, modifying it to include the limiting conditions under which we investigate the kinds of natural phenomena, yet without losing sight of the broader mechanistic context embodied in the meanings of nature-as-God-matter-and-form.

The far-reaching adjustments get underway as soon as Descartes begins to qualify his own "nothing-but" statements. Once in possession of his metaphysical foundations and his controlled advance into the philosophy of nature, he can safely interpret his flat declarations that "my entire physics is nothing else than geometry," that "my entire physics is nothing else than mechanics," and that "the rules of mechanics . . . are the same as those of nature."[27] These statements must undergo systematic modification when the concept of nature is being taken, not at the foundational level of *general* physics, but now in the more differentiated and *specialized* portions of physical study.

In the sixth chapter we have already noticed how Descartes initiates this adjustment with his concretizing of the mathematical component. The geometry needed for the study of nature is not an abstract but a concrete geometry, that is, one which is adapted to the mechanical study of movement and the varied structural phenomena in nature. His main point now is that the rules of mechanics, to which a concrete geometry is proportioned, do not themselves furnish an exhaustive description and explanation of the moving phenomena in the world of actual perception. The rules of mechanics enable us to formulate the primary and secondary laws of mechanical movement, and to this extent they are identified with the laws of nature. But what becomes specified thereby is only the general formal component of nature, corresponding to the general view of particulate and corpuscular matter. Although a knowledge of these aspects is necessarily involved in the further study of nature in its actual formations, it cannot satisfy our drive toward understanding the more determinate structures and relations among moving bodies in the experienced world of men. Because the meaning of nature is notably enlarged through the use of scientific aids beyond those furnished by

[27] *Letters of July 27, 1638, and April 30, 1639*, to Mersenne and Debeaune (*Correspondance*, vol. II, p. 363; Alquié, vol. II, p. 129); *Discourse on Method*, Pt. 5 (Olscamp, p. 44). Cartesian general mechanics admits only those reasons which are mathematically evident, so that Cartesian nature will do everything geometrically. In this sense, the limitations in Descartes' conception of a geometry of algebraic functions are visited upon his conception of a mechanicist philosophy of nature, preventing it from enjoying the mathematical-mensurational resourcefulness of the generation of Newton and Leibniz. This essential restriction upon the Cartesian effort to mathematicize and mechanicize nature is underlined by J. Vuillemin, *Mathématiques et métaphysique chez Descartes* (Paris, 1960), pp. 93–97, and is the heart of the contrast between Descartes' geometrism and Leibniz's analytic functionalism proposed by Y. Belaval, *Leibniz, critique de Descartes* (Paris, 1960).

the rules of mechanics in general physics, the Cartesian conception of nature cannot restrict itself to any reductive core of concrete geometry and mechanics.

Critics of Descartes have remarked that he does not actually employ mathematical concepts and notations to explain his theory of nature, and that he relies upon quite crudely conceived models and thought experiments to apply his mechanical principles to the visible world. He himself is well aware of these traits and directs his correspondents' attention toward them. But he regards them less as defects than as unavoidable consequences of the concretizing movement and communicational side of his theory of nature.[28] He draws a distinction between the moment of discovery in philosophy of nature (for which there must be a technical grasp of mathematics and mechanics, along with a disciplined use of imagination) and the work of communication to others. Except for the *Geometry* itself and parts of the *Optics*, none of his writings require the reader to make formal

[28] *Entretien avec Burman*, ed. by C. Adam, *op. cit.*, pp. 120–124. Comparing the three scientific treatises published along with his *Discourse on Method*, Descartes states that: (a) the *Geometry* is purely mathematical, the *Meteorology* is purely in philosophy of nature, and the *Optics* combines mathematics and natural philosophy; (b) the *Geometry* demonstrates in a rigorously mathematical way, whereas the intent of the *Optics* and *Meteorology* is rather to persuade readers about the general superiority of his method; (c) although when separately considered, the specialized parts of natural philosophy lead only to probability, they do have the convergent force of a physical (but not purely mathematical) demonstration when considered together. *Letters of May 1637, October 3, 1637, and December 1637*, to Mersenne and Plempius (Alquié, vol. I, pp. 540, 795, 820). Such convergent physical reasoning (which uses metaphysical and mechanical principles along with physical hypotheses and experiments) constitutes an appropriate substitute or *locum tenens* for a purely mathematical demonstration. This governs Descartes' carefully worded claim that, in passing from general physics to more specialized questions, he will "admit nothing about those [material things] as true, which is not derived so evidently from these common [geometrico-mechanical] notions that it must be counted for a mathematical demonstration," *pro mathematica demonstratione sit habendum. Principles of Philosophy*, Bk. II, sect. 64 (AG, p. 221). On the methodological mixture involved in this distinctive inquiry into particular physical issues, see G. Buchdahl, "Descartes's Anticipation of a 'Logic of Discovery' " in *Scientific Change*, ed. by A. C. Crombie (New York, 1963), pp. 399–417. That Descartes sets a pattern for the close adaptation between metaphysics, scientific methodology, and basic physical concepts in classical modern philosophy, is established in W. von Leyden's *Seventeenth-Century Metaphysics* (New York, 1968), and in G. Buchdahl's *Metaphysics and the Philosophy of Science: The Classical Origins, Descartes to Kant* (Cambridge, Mass., 1969). In terms of the analytic classification of philosophies of nature proposed by Ernan McMullin, "Philosophies of Nature," *The New Scholasticism*, vol. 43 (1969), pp. 29–74, the historical Descartes develops a PNM: a mixed type of philosophy of nature which draws upon both a broader epistemological-metaphysical position and also upon basic traits in the actual methodology of the empirical science of his time. This blending is urged polemically in Descartes, *Philosophical Letters* (K, pp. 33, 38–44).

use of mathematics or to follow all the research techniques used in adapting mathematics and mechanics to the study of the world of actual movements and bodily configurations. Yet in order to inform themselves sufficiently to give well-founded assent to his theory of nature, students must appreciate the broader spirit of mathematical thinking and respond to its demands.

Descartes furnishes a pioneer example for the entire line of rationalist philosophers by distinguishing between the restricted technical meaning of the mathematical disciplines and the broader philosophical meaning of mathematical thinking. Without this distinction, we cannot grasp the sense in which he recommends his philosophy of nature for its mathematical quality. It is a mathematically organized theory primarily because it gives reflective expression to the *ingenium mathematicum*, the basic mathematical impetus of the mind. Although this way of thinking displays itself quite remarkably in the formal mathematical disciplines, it does not confine itself to these modes of its activity. For in the last analysis, the mathematical spirit signifies nothing less than the true disclosure of the *ingenium humanum* itself: the mind of man operating in a peak manner for realizing its own capacities and moving toward the goal of unified knowledge or human wisdom.

Whenever he refers in this more capacious sense to the mathematical way of philosophizing about nature, Descartes characterizes it by four basic marks.[29] First, it consists in an *ability to distinguish* true and demonstrative reasoning from what is false or only probable. This may be done in terms of formal entailment from a set of definitions and axioms, as in geometry. But usually in philosophy of nature,

---

[29] See J. L. Allard, *Le Mathématisme de Descartes* (Ottawa, 1963). But in regarding the incomplete and hypothetical character of Descartes' special physics as a defect in the universalization of mathematical method (pp. 163–222), Allard does not allow sufficiently for the Cartesian transformation of the mathematical factor as functioning within natural philosophy. Descartes himself marks this thorough adaptation by progessively reinterpreting the meaning of "demonstration" in natural philosophy. (a) He criticizes those who restrict demonstration to abstract geometry; who demand this pure sort of demonstration in physics; and who overlook the criterion of the geometer's nondisagreement with experience, in additional to internal coherence. *Letter of May 17, 1638*, to Mersenne (Alquié, vol. II, pp. 62–64). (b) Descartes also distinguishes two appropriate modes of physical demonstration: the "explaining" of effects by showing how determinate physical phenomena can be conceived as following from specific hypotheses; and the "proving" of the hypothetical principles of particularized physical explanation themselves by establishing their verifying relationship with the experienced phenomena. *Discourse on Method*, Pt. 6 (Olscamp, pp. 60–61); *Letter of February 22, 1638*, to Vatier (Alquié, vol II, pp. 29–30); *Principles of Philosophy*, Bk. III, sects. 4, 42–47 (AG, pp. 223–226). See Descartes, *Philosophical Letters* (K, pp. 45–49, 55–56), for cautions on the mathematical claims for philosophy of nature.

one must consider the reality reference of basic propositions. The mathematically educated mind discerns that a philosophy of nature grounded in the Cogito weakens the ability of skepticism to cast doubt upon the existential bearing of concrete geometry and mechanics.

Second, the mathematical spirit exhibits itself in a settled capacity for *developing many consequences* from a few well-defined and established principles. Once more, however, the relationship of the-many-from-the-few diversifies itself in accord with the more restricted or the more comprehensive aim of the inquiring human mind. The variables increase notably when this relationship is specified as holding good for the reality principles governing the determinate theory of nature. For here, the inferred consequences must be complex enough to include the many kinds of judgments and passional dispositions in man and the many sorts of sensible phenomena in our world.

Next, the effective presence of the mathematical approach depends upon the degree in which there is a process of *continuous discovery and ordering* of truths, correlated in such fashion that they follow closely upon one another and constitute a unified body of knowledge. This continuous, organizing movement of the mind seeks to realize the ideal of concatenated reasoning and organic unity of certitudinal knowledge in all areas of philosophical inquiry. Yet it cannot run roughshod over the diversified actualities presented by our intellectual life, the bodily world, and the field of moral decision, together with the proportionately diversified instruments for probing into them.

As a final note of the mathematical mind, therefore, Descartes requires constant *fertility in devising new methodological tools* for exploring the different fields of study and coping with their distinctive problems. For although the living mind of man remains the same in its native resources and ultimate goal of wisdom, it must still adapt itself to new regions of reality and discover new intellectual means for building upon the foundational metaphysical truths about the Cogito and God. The need for conceptual and experimental innovations is specially acute when we are investigating the phenomena and structures in sensible nature.

Along with his programmatic insistence upon the ideal sameness of all acts of intuition and deduction, then, Descartes also recognizes the human need for methodological diversifications in the search after philosophical wisdom about natural reality. His appeal to the mathematical spirit in the study of nature does not signify that the world of nature must be submitted to a univocal mathematization and depersonalization of the inquiry into it. Instead, the Cartesian philosopher of nature seeks his own distinctive actualization of the four "mathematical" traits of indubitable certitude, development of many consequences from a few principles, continuity in systematic reasoning, and use of new intellectual instruments.

(1) The theory of nature's rooting in metaphysics assures him a criterion of truth, not for the easy resolution of every question, but precisely for assuring the ontic bearing, universal scope, and conditioned necessity of the basic laws of movement. Furthermore, the metaphysical preparation forearms the philosopher of nature against accepting either an uncritical realism of sense experience or an uncritical formalism of the mathematical view of reality. Thus a core of certitudinal knowledge belongs to his meaning of nature. Whether or not he will be able to achieve the same high degree of certitude in particular inquiries as he does in formulating the primary mechanical laws of nature will have to wait upon the outcome of the actual inquiry, rather than be stipulated beforehand.

(2) From the primary principles of movement, the mathematical philosopher of nature is indeed able to derive some secondary rules and suggest several more restricted means of interpreting sensible phenomena. But he also finds that he cannot simply derive the multifarious phenomena in nature through conceptual analysis of this small number of principles. Even though the general presence of such principles is supposed, new theories and imaginative aids must be devised in order to engage in actual investigation of the varieties of physical reality.

(3) The mathematically directed mind will make continuous transitions back and forth between the general mechanical principles and the more particular concepts and models. Yet if it is critically responsible, it will not permit the psychological unification to gloss over the different levels of evidence and certitude involved in the unifying process. As a safeguard against such self-illusion, Descartes seldom fails to include some explicit methodological references to the many "instruments of deduction" required to make the unifying transitions from one level of inquiry and region of visible reality to another.

(4) An internal study of the genesis of these instruments of deduction constitutes the fourth positive mark of the efficacious presence of the mathematical spirit in developing the philosophy of nature. Descartes gives an operational description of how the mathematically ordered mind works inductively and experimentally with the actual composites in experience, and develops the philosophy of nature by framing hypotheses and constructing imaginative models of the kinds of bodies in our world. Such methodological analysis is

a peculiarly effective way in which the Cartesian claim to know nature is tempered by the capacities of the human mind, even in its mathematically regulated operations. Hence it deserves our separate consideration in the following chapter.

9

Methodic Convergence upon Operable Nature

HUMAN reasoning in natural philosophy is methodologically complex and mixed. Descartes stresses this point directly in proportion to recognizing the importance of knowing the actual world with sufficient determinateness to gain some human mastery over it. His meaning of nature becomes increasingly modified by our practical purpose for the study of nature and by the diversification in methods imposed by the value of practicality. Thus his conception of nature is not purely speculative, impersonal, and value-indifferent.

This *motif of practicality* in the meaning of nature becomes prominent when that conception is affected by five aspects of our methodological adjustment to the world, considered as being not only actually existent, but also open to human operative control. These five means of convergence upon visible and operable nature are: hypotheses and experiments, sense perceptions, the use of models, moral certitude, and the modified ideal of a comprehensive mechanical knowledge of natural phenomena. Taken together, they specify and render palpable, as it were, the process of adjusting man's mathematical way of thinking to the practical task of studying nature and using it for his own welfare. Only a synoptic grasp of the following five methodological developments, taken both in their distinctness and in their operational unity, can do justice to what Descartes means by placing at human disposal his mathematico-mechanical philosophy of nature. For together, they constitute the disposing modifications in the meaning of nature which correspond to man's practical engagements with the world.

(1) It is clear to Descartes that even the theoretical development of scientific *hypotheses and experiments* arises purposively out of our human aim of obtaining pragmatic control over nature. Because we want to make ourselves somehow the masters of nature, we cannot remain content with the very general knowledge of it obtainable from the basic mechanical laws alone. They inform us about some common features of any world answering to the phenomenal description of our own, but they do not enable us to distinguish between

what could occur in any mechanical particulate world and what does actually obtain in our humanly experienced world. At the initial stage in the formation of the philosophical view of nature, it is true that such indifference to actuality constitutes no real drawback. But it does become defective as soon as we see the relevance, for a humanly adequate meaning of nature, of our technological efforts to understand and reshape that material reality wherein our needs arise and our satisfactions are met.

Here man's pragmatic engagements with the actual universe are paramount. Thus a concern to improve our practical relationship with encountered bodily things impels us to modify our working conception of nature with the help of hypotheses and experiments. For they relate the human mind more determinately, although with less certainty than do the general physical laws, to the existing world of experience and action.

Descartes states the theoretical aspects of this situation in a striking passage of first-person discourse, which underscores the inquirer's own aims and decisions in making a study of nature.

I reflected upon all the objects that ever presented themselves to my senses, and I venture to say that I never noticed a single thing about them which I could not explain quite conveniently through the [mechanical] principles I had discovered. But I must also confess that the power of nature is so ample and so vast, and these principles so simple and so general, that I almost never notice any particular effect such that I do not see right away that it can be derived from these principles in many different ways. And my greatest difficulty is usually to discover in which of these ways the effect is derived. And to do that I know no other expedient than again to search for certain experiments which are such that their result is not the same when we explain the effect by one hypothesis, as when we explain it by another.[30]

This text brings out both the strength and the limitation of the general meanings of nature. One can appeal with sufficient deter-

[30] *Discourse on Method*, Pt. 6 (Olscamp, p. 52). Descartes always eventually tempers his bold claims for a knowledge of nature by drawing attention to the many different instruments and limitations of the human mind engaged in the inquiry. Hence the actual Cartesian "deduction" used in the study of nature is a very mixed process. It must employ induction and enumeration, experiments and hypotheses, so that the human mind can achieve that specific blending of mechanical principles which will approximately "explain" and "prove" (see above, n. 29) the determinate modes of physical reality. Cf. *Rules for the Direction of the Mind*, Nos. 3, 6–8, and 11 (Lafleur, pp. 155, 164–179, 185–187); and L. J. Beck, *The Method of Descartes* (Oxford, 1952), pp. 111–146, 230–271. The methodological importance of the concluding sections of the *Discourse on Method* for modalizing both the procedural and the certitudinal aspects of Cartesian philosophy of nature is underlined by Elie Denissoff, *Descartes, premier théoricien de la physique mathématique* (Louvain, 1970).

minateness to the power of nature (taken as God and as moving particles of matter) and its mechanical laws or principles of movement, when there is question only of ruling out any non-mechanistic account of the movement and structure of bodies. The latter account would constitute an appeal to occult powers and qualities, that is, it would posit the presence in nature of aspects which are incompatible or unconnected with explanation through the laws of movement for extended particles.

But the very generality and fertility of nature, conceived at this level, open up alternate ways in which the actual bodies and relationships in our experienced world can be connected with the comprehensive principles of nature. What the quoted text leaves implicit is the reason *why* Descartes is dissatisfied with having a broad range of possible combinations for reaching the actual phenomena of our world. His dissatisfaction over the discrepancy is both theoretical and pragmatic. That situation is intolerable, both because it prevents us from acquiring a closely specified understanding of the formation and constitution of bodies, and because it thereby thwarts our practical interest in controlling nature for human advantage. Hence hypotheses are proposed and testing experiments devised in order to render our explanations sufficiently determinate, and our entire conception of nature sufficiently focused, for establishing that close understanding of the actual configurations and course of nature upon which our welfare depends.

(2) It will also be noticed that Descartes is prompted to supplement his broad mechanistic view of nature with specific hypotheses and experiments precisely in order to explain and control "all the objects that ever presented themselves to my senses," that is, the perceptual world of our experience and action. It is in *sense perceptions* that we become related with this experimental and actional reality which spurs us to the use of physical hypotheses and experiments. Hence one of the strongest motives underlying Descartes' cautious rehabilitation of the senses, in the latter portion of his metaphysics and his philosophy of nature, is the indispensable contribution they make to our pragmatic acquaintance with, and control over, material things.

He cannot remain content with either the old skeptical tropes concerning sensory illusions or with his own metaphysical critique of sensation. Such arguments establish no more than that we cannot place any primordial reliance upon sensory sources for our basic existential knowledge of ourselves and God, or for an essential grasp of extensive matter. Once sense perception is displaced from any foundational role in Cartesian metaphysics, however, its practical

F

importance can safely be acknowledged and its proper role in modifying the meaning of nature developed.

Methodically chastened sense perception helps to perfect our conception of nature in five closely related respects. First, it directs our experience toward the actually existent world, thus enabling the philosopher of nature to distinguish between the general logic of all mechanically ordered worlds and the determinate understanding of the actual course of nature in our world. Second, it presents us with definite composite bodies in the perceptual world and with differentiated structures and sequences among sensible phenomena. Only through this insistent testimony do we remain uncomfortably aware of the descriptive and explanatory inadequacies in the general mechanistic schema, and become concerned with the study of specific issues. Although sense perception does not itself formulate the specific problems and hypotheses, it provides the experiential acquaintance with *this and that* sort of bodily thing out of which the physical problems grow, and in terms of which the proposed hypotheses are tested. It is the constant goad which prods Descartes out of any complacency he may have felt about his general meanings of nature.

In the third place, there is some kind of observational reference always involved in determining the results of experiments. Sense perception cannot decide the issue by itself, since the judgment about a relation between hypotheses and confirming experiment is an intellectual operation. But the perceptual report is a weighty factor in determining whether the course of events in nature would be any different under one hypothesis rather than another. A fourth function of perceptual sensory experience is to keep us insistently informed about the distinctive conditions and needs of the human inquirer himself. Thus it never permits us to forget the decisive inclusion of our composite human nature—along with its feelings, passions, and practical activities—within the actual world of nature and our needy relationship with the rest of natural reality.

Finally, and by way of summing up these several contributions, Descartes appreciates the importance of properly ordered sense perception for constantly increasing the empirical relevance of his view of nature. It spurs on scientific investigators to develop specific theories and hypotheses for a better understanding of a perceived natural reality that is both vast in power and highly diversified in its actual structures and processes. And it reminds us that nature is not only quite determinate in its actual processes, but also quite *importunate* in its demands upon man. Sense perception forces us to measure the adequacy of a theory of nature at least in part by its usefulness for man, considered as a complex reality whose satisfactions are bound up with other determinate realities in nature. Thus the relevance which perceptual experience helps to assure for Cartesian thinking about nature involves a deliberate accommodation of its general meanings to some quite concretely humane and pragmatic relationships.

(3) The *use of models* is also essential to the Cartesian exploration of nature. Due to his acknowledged preference for geometry over arithmetic and his integration of concrete geometry with mechanics, Descartes favors the mechanical kind of models at various levels of inclusiveness. We have already heard him refer to his most general conception of nature-as-machine as "this great model," a designation which encourages students of nature to turn away from private, nonmechanical imagery and toward the inspection of a public structure. Here, the machine model functions on a grand scale as a cosmic metaphor for viewing all natural happenings. It guides us in applying universally the mechanical principles of intelligibility. Its impact upon the mind is to rule out as unreal and inappropriate any nonmechanical ("occult") types of general explanation of the course and structures of nature, and thus to set up a principle of consistency which must be honored by the more specific attempts at description and explanation.

In these latter inquiries, Descartes himself makes generous use of more restricted mechanical models. They represent a way of conceiving the determinate modes of convergence of mechanical forces, operating upon corpuscular particles in the definite patterns which form and determine the several kinds of bodies. The purpose of these specific models within Cartesian philosophy of nature is threefold. (a) They serve as a rough heuristic guide for thinking about the genesis and internal structure of many kinds of natural phenomena found in our experience; (b) they build a bridge of knowledge and reality between broad mechanical principles and special physical theories on the one hand and the perceptually appearing world on the other; and (c) they help us to conceive the infraperceptual domain of material particles. These three uses are closely bound together and, in the actual working out of the Cartesian approach to nature, they tend to form a blurred union.

Descartes himself is unevenly successful in distinguishing these functions and respecting their epistemological limitations. (a) He is most prolific and also least rigorous in proposing descriptive and explanatory models of the different sorts of bodies we encounter.

Usually, his proposals consist of verbal models and diagrams loosely sketched, rather than a careful construction of determinate mechanical factors. The verbal sketches enable him to think more concretely about the formation of snow and other meteorological phenomena. Descartes' own imagination is both stimulated and disciplined by working within the bounds set by mechanical models of the growth, internal pattern, and interworking of familiar sensible objects. At the same time, his models are easily communicable to others and help to render his mechanistic story of our world more convincing, through everyday applications.

(b) Yet in his books and correspondence, Descartes sometimes becomes critical of vivid models whose sole recommendation lies in the ease with which they can be imagined and communicated. Hence he requires of specific physical models that they show their determinate consistency with the general principles of mechanical explanation and the general meanings of nature. Furthermore, the models must be capable, at least in principle, of serving as explanatory mediators in the total movement toward a more determinate understanding of the perceptual actuality of nature. They have to contribute something toward the reality reference of scientific inquiry, as it seeks to join the predictive hypotheses more definitely with the sense observations used in the confirming process.

Such involvement of models in the confirming operation has the important epistemological consequence of permitting Descartes to acknowledge a legitimate role, in the theory of nature, for methodically controlled probability and responsible conjecture. Just as he rehabilitates sense perception in connection with the use of hypotheses and confirming experiments, so does he accept a disciplined use of imagination and conjectural or probable reasoning in the construction and use of models. Thus the Cartesian meaning of nature is not confined methodologically and epistemologically to what can be attained with demonstrative certitude, but includes those aspects which we can grasp only through perception and imagination, conjecture and probable inference.

(c) That these exploratory and analogical uses of the mind make more than a superficial addition to the theory of nature is clear from the third function assigned to models. Descartes never anticipates the mathematical resources of Newton and Leibniz in measuring the forces among the basic particles of matter. Yet he does not leave this problem entirely unprobed, since one primary purpose of mechanical models is to supply an indirect route for understanding and perhaps controlling the sensibly imperceivable particles and their force patterns. Both in his early years and in maturity, Descartes recommends the use of macroscopic models, embodying mechanical principles, as a basis of analogy for grasping something about the submicroscopic structures and forces in nature.

His typical declarations on this issue deeply impressed the pre-Newtonian generation of British scientists and philosophers, especially Boyle, Hooke, Sydenham and Locke himself.

> Knowing that nature always acts by means which are the easiest and simplest of all, perhaps you will not judge that it is possible to find there [in the infrasensible order] anything more like those things of which she [nature] makes use than those which are here proposed. . . . There is nothing more conformed to reason than to judge about things which, because of their smallness, cannot be perceived by the senses, by the example and on the model of those we do see. . . . So, just as men with experience of machinery, when they know what a machine is for, and can see part of it, can readily form a conjecture about the way its unseen parts are fashioned; in the same way, starting from sensible effects and sensible parts of bodies, I have tried to investigate the insensible causes and particles underlying them.[31]

Each of these statements contains interesting implications for Descartes' effort to concretize his conception of nature.

In the first one, he appeals to the general traits of nature's simplicity, self-consistency, and economy in the use of materials and means for shaping the experienced world. But neither his metaphysics of God nor his reflective methodology will permit him to follow Galileo in treating these traits as coercive first principles for determining how the components in nature *must* unconditionally be structured and related. Hence he makes a modest judgment based on the greater *likelihood of there being an analogy*, rather than utter disparateness, between the sensible and the infrasensible aspects of nature. Descartes' rather cumbersome way of stating this likelihood only emphasizes the situation of working through appropriate

---

[31] *L'Homme* (Alquié, vol. I, p. 479); *Letter of October 3, 1637*, to Plempius (Alquié, vol. I, p. 793); *Principles of Philosophy*, Bk. IV, sect. 203 (AG, pp. 236–237). On the use of diagrams, sketches, and plates, as visual and imaginative aids to the understanding in natural philosophy, see *Rules for the Direction of the Mind*, Nos. 12, 14 (Lafleur, pp. 192–193, 211–222); *Letter of July 13, 1638*, to Morin (Alquié, vol. II, pp. 75–76); *Entretien avec Burman*, ed. by C. Adam, *op. cit.*, p. 122; and, of course, the actual illustrations included in the scientific treatises. The limitations of his mathematical tools conspired with Descartes' liking for easily apprehended thought-experiments in physics to induce him to substitute loose analogies for exact measurement, although he deplored easy imaginings which are *nowise* disciplined by mathematico-mechanical thinking. M. B. Hesse, *Models and Analogies in Science* (Notre Dame, 1966), brings out the current theoretical interest in such issues.

conjectural models, which do not yield an intuitive envisagement of the unseen structures and processes in nature.

That this is a question of reasonableness of judgment about the probably real significance of models, rather than a necessary conceptual understanding, comes out unmistakably in the second statement. There, the quite indispensable role of sense perception in forming mechanical models is recognized. In this type of model formation, the controlling analogies are drawn from the perceptual world, and hence Cartesian philosophy of nature demands the epistemological recovery of nature qualified in terms of visual perception. To establish some continuity of meaning between the very general senses of nature and the insistent perceptual world, Descartes finds it economical and reasonable to suppose that the particles and forces in the infrasensible order are not entirely unlike the mechanically describable parts of larger bodies. The chief difference is the (for him) noncritical one of size. Insofar as models help to specify this similarity, they yield both an analogous apprehension of the infraperceptual order in nature and a sort of mechanical measurement or working guide for our further research.

The final sentence in the quoted text reveals that the rough and approximate character of such mechanical frameworks, through the analogical use of models, does not escape Descartes himself. He seldom fails to restrict the knowledge claim for his philosophy of nature, at this critical phase in its development. He realizes that we sometimes find that machines similar in outer appearance, nevertheless, have quite dissimilar internal parts. Thus his mechanical models will yield a conjecture or likely pattern, without pretending to give direct essential insight into the basic particles and forces in nature. Epistemologically, Descartes (as distinct from his more incautious followers) is content to determine what may well be the particulate structure and mechanical genesis of experienced bodies, even though perchance they are not actually of this sort (*etsi forte non talia sint*). He accepts the qualified aim of showing how natural bodies could be made in terms of his mechanical models of description and explanation, although one ought not to conclude that they are really made in this fashion (*non tamen ideo concludi debet, ipsas revera sic factas esse*).[32]

---

[32] *Principles of Philosophy*, Bk. IV, sect. 204 (AG, p. 237). On absolute and moral certitude in philosophy of nature, see *ibid.*, Bk. IV, sects. 205–206 (AG, pp. 237–238). This Cartesian distinction must be kept apart from the conventional threefold division into metaphysical, physical, and moral kinds of certitude. For Descartes, there is absolute certitude both in metaphysics and in the general principles of physics; similarly, there is moral certitude sufficiently definite for directing our action both in the specialized parts of physical explanation and in morality.

Thus the Cartesian theory of nature includes a clear recognition of the limits of human intelligence in respect to the more determinate modes of active interplay (*concursus*) of mechanical forces and particles, when they join in constituting the concrete bodies and visible configurations in our world. Yet the resultant conception of nature is not a sheer fiction, since it develops within a broader mechanistic and metaphysical context and does enhance our practical attitude of controlling and reshaping the particular materials and agencies in the world. Descartes agrees on this point with Francis Bacon that, measured by the yardstick of encouraging human practical control over nature, the approximative analogies obtained through mechanical models are as good as the results obtainable through perfect insight and demonstrative inference about nature's workings.

(4) It is through his *theme of moral certitude* that Descartes avoids the excesses of both a skeptical fictionalism and a metaphysical absolutism in regard to the kind of knowledge obtained in philosophy of nature. Although he avoids the former extreme through his metaphysical and mechanical principles taken in conjunction, he also refrains from extending the absolute certitude of these foundational principles to the entire investigation of nature. Taken as a whole, the several kinds of instruments and inquiries involved in developing our understanding of nature insure that it will be a complex synthesis of absolute and moral certitudes, rather than a homogeneous kind. The more we incorporate specific models and analyses of our experienced world into the meaning of nature, the more does our determinate meaning consist of moral certitudes, which enjoy a relatively high probability and practical importance.

Descartes requires that any candidate for inclusion in his detailed physical theory must meet three minimal criteria for determinate physical knowledge. (a) It must be formulated as an internally coherent signification; (b) it has to show consistency both with the general mechanical principles and with all the available phenomena of sensible nature; and (c) it must exhibit a practical relevance for the slow human work of observing and hypothesizing, experimenting and gaining pragmatic control over the environing world. In meeting these criteria, however, the proposed theory also fulfills the requirements for attaining moral certitude in physical matters. When all these marks concur to enhance our understanding and use of the perceptual world for human purposes, then the philosopher of nature has moral certitude about the soundness of this particular component in his view of nature.

Anticipating a favorite metaphor of Boyle, Schopenhauer and

recent geneticists, Descartes sometimes envisages the philosopher of nature as a cryptographer who is trying to break the code of nature and render it available in a coherent message. The possibility of there being an entirely different meaning for the determinate phenomena of nature is not entirely ruled out, as it is at the level of the absolute certitude we have concerning metaphysical doctrines on God and the Cogito and concerning the mathematico-mechanical basic laws of nature. Yet many people are unable to follow the fundamental Cartesian reasoning in metaphysics and mechanics, and hence they may suppose that the principles in this philosophy of nature are held only by chance and without any evidently reasonable grounds.

It is to persuade such incompletely formed minds that Descartes suggests that nature be viewed as a giant cryptograph, and his philosophy of nature in its specific prolongations as a strenuous decoding thereof. "Perchance they may acknowledge, however, that it could scarcely [*vix*] happen that so many things cohere together, if they [the principles of physical explanation in determinate nature] were false."[33] The adverb *scarcely* remains in force as a cautious qualification attached to the deciphered message of nature at the level of the humanly environing world. Descartes intends it to indicate that his determinately specified philosophy of nature does not convey pure and absolute certitudes. But it does furnish sufficient moral or working certitudes to constantly increase our theoretical and practical grasp of the visible universe. Having previously laid a foundation of metaphysical truths and absolute certitude against skepticism, Descartes now feels free to admit a large sphere for the pragmatic considerations of human living within natural reality. Thus his conception of nature is duly enlarged to include not only the absolute certitudes embodied in the primary mechanical principles, but also the moral certitudes required for dealing with the natural world in its capacities for satisfying human needs.

[33] *Ibid.*, Bk. IV, sect. 205 (AG, p. 238). See *ibid.*, Bk. II, sect. 64 and Bk. III, sect. 1 (AG, pp. 221–222), on explaining-all-the-phenomena-of-nature. Thus a physical hypothesis attains moral certitude in the degree that it unifies the many phenomena of nature in the coherent, comprehensive, and illuminating manner required for understanding and using nature for human aims. To judge when a proposed hypothesis does have this illuminating capacity, physical investigators are advised by Descartes to observe weaving activities and arithmetical games, so that they may "become accustomed to penetrate each time, by open and recognized paths and almost as in a game [*quasi ludentes*], to the inner truth of things." *Rules for the Direction of the Mind*, No. 10 (Lafleur, p. 183). Moral certitude in natural philosophy depends upon gaining sagacity of interpretative judgment, which grasps the relationship between our hypotheses and the course of natural phenomena. The reference to play acknowledges the role of imaginative hypothesizing and experimenting in the specializ ed fields.

(5) Having observed how the convergence toward operable nature involves the epistemological qualifications introduced by the use of hypotheses, models, and moral certitudes, we are now in a better position to understand Descartes' reiterated ideal intention of explicating or *explaining-all-the-phenomena-of-nature* (*omnia naturae phaenomena explicare*). This hyphenated expression has a technical meaning in his writings which does not revoke the above modifications in his conception of nature, but rather brings them to a striking focus. Descartes often recommends his philosophy of nature as giving a comprehensive mechanical knowledge of sensible things which omits nothing within our perceptual range, thus enabling us to grasp the true nature (*vera natura*) of this visible world of ours.

The key words in this vaulting claim are "comprehensive," "omits nothing," and "true nature." The Cartesian philosophy of nature is comprehensive, not in the sense of giving exhaustive essential understanding of every kind of natural entity, but in the sense that its metaphysical and mechanical first principles have a universal and necessary scope. They constitute an unrestricted criterion of consistency in meaning, with which all particular concepts and hypotheses about nature must manifest their agreement. Nothing is omitted in principle, either by way of allowing some "occult" or privileged exception in nature to these general laws or by way of excluding mechanistic explanation from the study of man, organisms, or any other aspect of the perceptual world. Whatever the mode of the sensible phenomena, it can be legitimately explored, illuminated, and controlled by means of the mechanistic hypotheses and models used commonly for the description and explanation of nature.

As for the precise sense in which the inquiring mind is thereby endowed with a conception of the true nature of things, it can best be appreciated by comparing Descartes' attitude with that of those contemporary methodologists who did not follow his mechanical way, but who nevertheless influenced his ideal of explaining the phenomena of nature. One such source was the Czech educational reformer, Comenius, whose exalted program is restated by Descartes in this serpentine sentence:

> Just as God is one and has created nature [to be] one, simple, continuous, everywhere self-coherent and self-corresponding, consisting of a very small number of principles and elements, whence he has drawn forth an almost infinite number of things, but in three kingdoms (mineral, vegetable, and animal), mutually distinguished by order and grades: similarly, the knowledge of these things must be (in likeness to the one creator and the one nature) unique, simple, continuous, uninterrupted, resting upon a few principles (even on a single primary principle),

whence all the rest right down to the most specific are permanently drawn in an undivided linkage and most wise order, so that thus our contemplation of things as a whole and in particular may resemble a painting or a mirror, which represents very exactly the image of the universe and of its particular details.[34]

Descartes feels a nostalgic attraction toward this comparison, which takes the divine knowledge of nature as our exemplar and which likens our mind to a perfect mirror image of the cosmos, taken as a whole and in its particular modes. Nevertheless, he is careful to distinguish the Comenian educational project from his own claims for the philosophy of nature.

The difference lies in the fact that Descartes takes the epistemological question seriously. He includes the human manner of knowing in the very meaning for nature, and therefore cannot square his own findings about our knowledge of nature with the ideal of Comenius. In this strong sensitivity toward methodological and epistemological findings about man the knower and agent, Descartes resists taking a purely impersonal view of nature and of our relationship with it. The human mind is simply *not* related to nature as a mirror is related to a fully given object.

The mind itself is not a passive reflector of the order of nature,

---

[34] *Letter of 1639*, perhaps to Mersenne or Hogelande (Alquié, vol II, pp. 154–155); Alquié questions the authenticity of this letter on grounds of thought, style, and vocabulary, but it is expressing the Comenian position. See *The Great Didactic of John Amos Comenius*, tr. by M. W. Keating (2 vols., London, 1921), vol. II, pp. 98–202, on conforming the educational process to the total structure of nature. Descartes criticizes Comenius for dreaming that we can easily develop a universal science of nature, compress it into a textbook, and teach it to young students, as well as for determining physical issues by appeal to the Bible and revealed truths. *Letter of August 1638*, perhaps to Hogelande (Alquié, vol. II, pp. 81–82). For his own part, Descartes essentially qualifies his ideal of understanding all the phenomena of nature—*omnia naturae phaenomena, tous les phénomènes de la nature*—by also riveting our attention upon the limits of the human mind and its methods. In our reflective conception of nature, we must include some reference to these limits and to the proportionate methodological devices (induction, hypothesis, and analogy) and state of knowledge (moral certitude) involved in grasping the *vera natura* of physical reality as a whole. *Principles of Philosophy*, Bk. II, sect. 64; Bk. III, sect. 42; Bk. IV, sect. 199 (AG, pp. 221, 223, 234); *Letter of November 13, 1629*, to Mersenne (Alquié, vol. I, p. 226); *Letter of June 1645* (*Correspondance*, vol. VI, p. 242). With the later history of science chiefly in mind, E. J. Dijksterhuis, *The Mechanization of the World Picture* (Oxford, 1961), pp. 408–409, judges that Descartes' own goal is an illusory but fruitful ideal, bringing mechanicism beyond Galileo's piecemeal approach to nature and thus preparing for the Newtonian natural philosophy. Chapter 3 of Robert McRae's *The Problem of the Unity of the Sciences: Bacon to Kant* (Toronto, 1961), weighs Descartes' contribution to the ideal of unified science. For a critique of Comenius, cf., Descartes, *Philosophical Letters* (K, pp. 59–61).

since the entire Cartesian analysis of the process of philosophizing about nature stresses our need for initiative and effort. We must strive constantly to devise and improve the concepts, models, and techniques for studying the world and making it serve humane ends. Our attitude is not that of spectators gazing in a mirror, but that of active investigators and experimenters, users and enjoyers of natural reality. Here, Descartes agrees with Francis Bacon's remark that only the angels can afford to be onlookers at nature's processes: men can grasp and utilize these processes only through their active efforts of research.

The Cartesian knowledge of nature is never perfectly established and rounded off, like a painting waiting only for the artist's signature. Rather, there is a complex intermixing of the several levels of principles and theories, certitudes and hypotheses, all serving to advance another step in our continuous interpretation of natural phenomena. To-explain-all-the-phenomena-in-nature is Descartes' compendious way of synthesizing the developing relationship between the many factors he has distinguished in the knowledge of nature. This ideal stands for an open process of weaving together the evident principles of metaphysics and general physics, the reports and hypotheses on sensory phenomena, the mechanical models and practical usages, the absolute and moral certitudes.

Only in this synthesizing process as a whole does Descartes locate his mathematical demonstration and mechanical knowledge of nature. And hence only here does he find the *vera natura* of the visible world, insofar as it is humanly attainable. To be sure, the result falls far short of an exhaustive Comenian contemplation of the universe. But Descartes recommends this explicating synthesis as an appropriate mode of human philosophical wisdom and as a spur for enlarging man's practical control over nature. What renders the Cartesian conception of nature unmistakably distinctive is not a superadded signature, but the inherent style of developing and interweaving the components in the theory of the synthesized phenomena of nature.

10

Natural Belief and Humane Finality in Nature

IT would betray the spirit of Cartesian philosophy of nature to overlook the reflective dimension of the problem, namely, the self-inclusion of man in the order of nature. It is not enough to require a mechanistic explanation of the living human body, and then to point out that the thinking self or mind is substantially distinct from the entire order of bodies. Descartes recognizes that there is something unique and existentially irreducible about the union of active composition between the human body and the human thinking self. So basic is the real unity of the human composite that its meaning cannot be reconstructed from his separate metaphysical analyses of mind and body, but requires some further interpretation of the presence of man within nature. Hence Descartes permits his probing of the meaning of nature to exert a retroactive effect upon the heart of his metaphysical reflections on man.

As a sign that his theory of nature does not merely flow in a one-way direction outward from his metaphysics, but raises issues which reinvade metaphysical territory and require further consideration, the *Meditations on First Philosophy* does not conclude on a note of dualistic triumphalism separating mind and body. Rather, this basic metaphysical treatise reaches its climax in a very complex discussion of the passage from the perspective of the *separate Cogito-self-mind* to that of the *composite unity of man*. Central to this transition, and making it unavoidable for a responsible philosopher of nature to reconsider, are two problems in the epistemology of nature which have metaphysical repercussions. They concern the meaning and legitimacy of appeals which people sometimes make to beliefs toward which "nature" inclines them, as well as to purposes which "nature" seeks to achieve. For understanding and evaluating these specific meanings of nature in terms of belief and finality, Descartes finds it necessary to study the relationship between the minded self and the man, or the concrete and composite human reality. Our interest in the question centers upon how this issue illuminates the Cartesian signification for the theme of nature and the natural.

Descartes describes how the ordinary fallibility of human life can generate an epistemological crisis concerning the trustworthiness of every normative appeal to nature. Relying upon our habitual inclinations (which we refer to as natural principles of belief), we are often led astray about the qualities of things and their contribution to our welfare. In defense of the basic reliability of natural beliefs, we assign mistaken sensations and harmful desires to an unhealthy condition of the human organism. We may attribute a misjudgment about color to a person's jaundiced condition, or an excessive desire for liquid to his dropsical state. Still, it is salutary to bear in mind that even these pathological conditions are thoroughly natural phenomena. As its meaning converges upon the human organism, nature includes both the healthy and the sick aspects. It is difficult to determine which aspect of nature is responsible for any particular belief that a man may strongly entertain on the grounds of being inclined naturally to hold it. In Montaigne's cautionary metaphor, the garden of nature brings forth our stubbornest convictions as so many cabbages in a very mixed lot, all of which are destined eventually to rot and betray anyone who depends uncritically upon them for understanding and sustenance.

Descartes realizes that his task is not to rule out wholesale all talk about natural beliefs but, instead, to sift and analyze such talk for whatever clues it can yield for determining more exactly how his conception of nature is modified by explicit reference to judging and acting men. Hence he launches into an extensive linguistic and epistemological analysis of the statement ordinarily made to justify our belief that ideas of sensible things are drawn from the things in physical nature and resemble these sources. When challenged about this assumption, we usually give this as our reason: "Indeed I seem taught thus by nature."[35] But this is not so straightforward a declaration as it may appear to be at first glance.

---

[35] *Meditations*, Bk. III. The Lafleur reading is: "It seems to me that nature teaches me so" (p. 95). This follows closely the French translation: "*il me semble que cela m'est enseigné par la nature.*" (Alquié, vol. II, p. 435: "it seems to me that that [our belief in the resemblance of adventitious ideas to their objects] is being taught me by nature.") The Latin text reads: "Nempe ita videor doctus a natura." (Alquié, vol. II, p. 194: "Indeed I seem taught thus by nature.") The skeptical interpretation of the relation between our ideas and nature's reality is expressed in Montaigne's persuasive vegetal metaphor: "If nature enfolds within the bounds of her ordinary progress, like all other things, also the beliefs, judgments, and opinions of men; if they have their rotation, their season, their birth, their death, like cabbages; if heaven moves and rolls them at its will, what magisterial and permanent authority are we attributing to them?" *Essays*, Bk. II, sect. 12 (tr. Frame, p. 433). Descartes replies that nevertheless the human mind also embodies an active impulse toward wisdom, can exercise reflective caution in sifting the natural propensities to believe, and can control its reality judgments

Indeed, Descartes finds it necessary to unravel at some length the complications concealed in this statement. The main difficulty stems from the polyvalence of the key terms: "nature" and "I," as they are used in the statement. Their several senses have to be sorted out, and the different meanings ascertained for their justifying use in the context. Only in the course of such analysis can Descartes establish the precise import and the appropriateness of the main verb "seem." He seeks to show the grounds for methodically using that verb in qualification of *all* discourse about beliefs purportedly resting on the trustworthy promptings of nature in man.

We are already familiar with the general meanings of nature as God-matter-mechanical-laws, but we have also seen that they do not provide the sufficient principles for understanding the more determinate modes and existential situations of natural beings. The present problem is to specify the meanings of nature still further, in order to render them more adequate to the complex relationships holding between man and the rest of nature. Such relationships enter prominently into practical human discourse, as the statement under inspection testifies. There is practical urgency, as well as theoretical interest, in clarifying those senses of nature which cluster around the active interchanges between men and the perceptual world.

Descartes proposes three more restricted meanings of nature, therefore, in order to obtain an understanding of how the dynamic concursus, or coordination of components, in the universe converges concretely upon the reality designated as *men-in-nature*. Depending upon the perspective and precise intent of the human inquirer and speaker, nature as the active system or assemblage of things signifies: (a) the formation and constitution of bodily things alone; (b) the realm of minds engaged in the search for wisdom; and (c) the union of composition between body and mind in man. Since we have already examined the epistemological complications entailed by a study of nature in sense (a), as constituting the world of various

---

by distinguishing different meanings of nature. Hence he specifies these five pertinent meanings of nature: (a) God as the author and conserver of the world system; (b) the entire internal coordination of things in their material components; and formal laws; (c) this dynamic order considered precisely in its disposition toward the bodily world alone or toward the thinking self alone; (d) the unified coordination of things as bearing precisely upon the human composite of mind and body together ("the man"); and (e) more or less adequately founded beliefs and judgments about a thing's conformity with some human concept or standard. *Meditations*, Bk. VI (Lafleur, pp. 130–143); *Replies to Objections*, No. 4 (HR, vol. II, pp. 106–107). See M. Guéroult, *Descartes selon l'ordre des raisons* (2 vols., Paris, 1953), vol. II, pp. 159–163; and L. J. Beck, *The Metaphysics of Descartes* (Oxford, 1965), pp. 257–261.

kinds of bodies, we can now concentrate upon the two latter meanings.

The statement under analysis is difficult to interpret, precisely because it involves fluctuation between senses (b) and (c) of the more particularized meanings of nature. To make us aware of this ambiguity, Descartes makes an even more effective technical use of the demonstrative adjective here than in his earlier account of fabular nature. His telling references to "this nature" and "that nature" (*haec natura* and *ista natura, cette nature* and *telle nature*) compel us to take into account not only the passage from the general to the more restricted and pluralized meanings of nature, but also the fluctuation of our intention between signifying now the mind's act of true judgment (nature in meaning *b*) and now the human composite itself (nature in meaning *c*). Clarification of the sense in which nature seems to teach us something or to incline us toward some belief depends upon our achieving some reflective control over the this-or-that type of predication of nature.

Yet Descartes realizes that these two latter particularized senses of nature are essentially correlated with different perspectives taken on man's own reality. Hence a methodic control over the this-or-that usage of nature requires a reflective understanding of the corresponding meaning of "I," in the key statement. The "I" who is said to receive instruction and support from nature, in the forming and holding of beliefs, can be construed in two ways. It signifies either *the thinking self* (the mind as reflectively aware of its own distinctive reality and evidential criteria—corresponding to nature in meaning *b*), or else *the entire man* (the union of composition between mind and body—corresponding to nature in meaning *c*). Hence any assertion that the rest of nature seems to be adapted to "me," in function of natural beliefs, stands in need of a fundamental clarification concerning the precise "I" in question. The active coordination of natural factors for determining beliefs about the actual world must be specified and differentiated, insofar as the other pole of the relationship will be either the thinking self or the entire man, in the Cartesian sense of these terms. Since the cognitional and actional requirements of the self and the man differ, there is a significant difference in the kind of appeal that can be made in each instance to natural belief and natural desire.

Nature-as-correlated-with-the-man displays itself in our human sensations and feelings, in our passional inclinations to seek or avoid things in the world, but it does so always in terms of meeting our practical needs and desires. Hence the beliefs about the sensible world and its qualities which express the proportioning between natural processes and the human composite are pragmatically grounded, and help us to determine what is useful and harmful for our daily living. There is some cognitive worth in such beliefs, but only when their limited reference and intention are carefully weighed. Convictions generated in this proportionment are always vulnerable to sensory illusions, distortions induced by ill health, prejudices of the everyday viewpoint, and the beclouding or precipitate effect of the passions upon our judgment.

By speaking compoundly and awkwardly about nature insofar as it concerns the needs of the man, Descartes hopes to put us on guard against these shortcomings from the standpoint of sound knowledge and truth claims. But the insistent danger is that we will convert a pragmatic rule of living into an infallible rule of knowing. We are inclined to overlook the human pressures and infirmities that help to determine our natural beliefs about the world and ourselves, thus placing more noetic reliance upon nature in this sense than is warranted by the relationship in question.

Yet we are not the supine victims of our natural convictions, so interpreted, but have a remedy in our free ability to regard the complexes of nature from the standpoint of the thinking mind or the critically alerted self. *Nature-as-correlated-with-the-self* is a basic tool of Descartes' epistemological criticism and philosophical reconstruction. For this complex meaning has an operational requirement. It emerges only as a consequence of submitting every spontaneous conviction and conventional belief about ourselves and the world to the process of doubt and evidential testing, as is required by the critical or reflective self. As long as we respect the methodological conditions for actuating this sense of nature and of naturally grounded convictions, we can attain some certain and well-founded knowledge.

An essential condition for grasping this compound meaning of nature is the resolve to forego any direct transition from our pragmatically generated sensations, feelings, and passions to our assertions of the truth about natural realities. The same epistemologico-metaphysical path laid out by Descartes for attaining true judgments in every other area must be followed by anyone seeking to understand nature-as-correlated-with-the-self. This is the most pertinent sense in which the philosophy of nature demands the mediation of metaphysics: to establish the reflectively controlled perspective of the wisdom-seeking self, to whose standard of existential truth all beliefs claiming to express our natural tendencies must submit. Only *after* such scrutiny is made, can the reflective meaning of nature function reliably in our judgments of reality.

To reinforce the critical testing phase in this meaning of nature, Descartes counsels us to say deliberately that nature "seems" to teach us such and such a view. This linguistic caution does not rest on an illusional or as-if notion of nature. It simply reflects the watchful methodic procedure of interpreting natural beliefs, first in terms of their pragmatic convenience for man in his complex needs, and only then in terms of the reflective self's criteria for true judgments of reality.

Descartes uses the term "seems" as a dispositional verb for submitting every appeal to a natural backing for our statements to a twofold principle of interpretation. For the *cognitive naturality* of our beliefs signifies either their utilitarian aspect (*utilitas ad vitam*), or their truth import (*cognitio ipsius veritatis*), in respect to the matter asserted. By systematically referring nature's effect upon the inquirer to this double test, we have it in our power to determine a this-or-that meaning of nature which will indeed deliver mankind from the tyranny of Montaigne's cabbage patch. For we contract to distinguish between the pervasively useful qualities of some convictions generated by man's senses, passions, and needs, and the noetic reliability of these or any other judgments insofar as the reflective self intends them to assert the truth about natural realities. The presence of utilitarian traits will neither disqualify nor directly guarantee the truth value of a position, since it must be further weighed in accord with the Cartesian criteria of knowledge.

This reflexive meaning of nature as cognitive naturality underlies the further distinction which Descartes drew between natural belief and the natural light.[36] He was prompted to distinguish between them for two reasons: because his own critics urged that he appealed too indiscriminately to the certainty of the natural light in support of his basic principles, and because of a confusion which he himself felt to exist between the two modes of cognitive naturality in the writings

---

[36] *Meditations*, Bk. III (Lafleur, p. 95); *Replies to Objections*, No. 4 (HR, vol. II, pp. 109–111); *Letter of October 16, 1639*, to Mersenne (Alquié, vol. II, pp. 143–146); and *Principles of Philosophy*, Bk. III, sect. 44 (AG, p. 224, for the *utilitas/veritas* distinction). For Descartes' use of *vis spiritualis* and the other ways of more carefully naming the human spirit, see *Rules for the Direction of the Mind*, No. 12 (Lafleur, pp. 191–192). On how the Cartesian image of nature is correlated now pragmatically with the human composite, its beliefs and desires, and now critically with the thinking self and its reflective judgment, cf. A. G. Balz, *Descartes and the Modern Mind* (New Haven, 1952), pp. 249–420. Descartes sharpens the distinction between "the light of nature" and Herbert of Cherbury's appeal to "universal consent" and an uncritical, undifferentiated "natural instinct," in *Philosophical Letters* (K, pp. 65–67). See F. Broadie, *An Approach to Descartes' 'Meditations'* (New York, 1970), pp. 60, 173–176.

of Herbert of Cherbury and other proponents of a native instinct for the truth. Men do experience a strong impetus to believe, in basic matters concerning their preservation and well-being. Such *natural belief* or impulse-to-believe does not directly concern itself with speculatively warranted truth, however, or furnish its own guarantee of certainty. Its truth significance for the study of reality can be determined only indirectly, by reinterpreting it within an independently established context of metaphysical principles for cognitive naturality in its epistemic meaning.

The paradigm case is provided by our natural belief in the existence of the material extended world. Within an intellectual climate permeated by skeptical questioning, it would be futile as well as philosophically confusing to appeal simply to the practical urgency and strength of this conviction—marks which could be admitted readily by skeptics, without conceding to it anything more than a high working probability. To determine the metaphysical significance of this belief, Descartes distinguishes between its practical importance and probability, and its function as a true deliverance about reality. To attain the latter aspect, he fixes analytically upon the unavoidable character of the belief and then interprets that unavoidability in the light of his previously established theory of God. A good and true God would not build into human nature such an incorrigible belief in the world as existent, if it were untrue. By incorporating our conviction about the existing material world within the context of metaphysical truths about God's veracity and creative production of the human composite, he also furnishes the indirect grounds for giving a certitudinal assent to the reality assertion conveyed through the natural belief of men.

Whereas natural belief concerns the cognitive dispositions of men as composite agents, the *natural light* signifies the vigor and telic activity of the self in its own actuality, considered both as having a divine origin and as striving toward wisdom. To express this correlation between the natural light and the striving self, Descartes employs a distinctive family of words. He calls the self a spiritual force and effort (*vis spiritualis, conatus*), a vivifying mind and seed of intellectual good (*ingenium, bona mens*). Our inner selfhood consists in a perduring effort toward making evidence more manifest, ordered, and comprehensive so that the perspicuous mind can assent and attain to the truth in an ever more adequate fashion.

By the natural light is meant, then, this dynamic inclination toward improving our grasp of the truth about reality, an inclination which may indeed be weakened and diverted by sense prejudices and the pressures of utility and convention, but which cannot be entirely quenched without destroying the self in its central act. Descartes does not regard the natural light as an esoteric criterion or as a cover for unexamined dogmatisms. For it is the nuclear principle of intellectual integrity and growth in us all. It signifies our own best self, responsibly inclining us through its own impulsion to submit our convictions to methodic criticism and a philosophical reconstruction leading to wisdom.

There is not a simple parity of indifference between natural belief and the natural light, therefore, considered as modes of cognitive naturality and as elucidations of the human meanings of nature. The former mode is always ordered to the intellectual requirements of the latter. Wherever our quest for well-founded knowledge of reality is involved, convictions generated through natural belief must be subjected to reinterpretation by those contextual principles of inquiry which survive the evidential probing to which our mind is inclined by the natural light itself. Precisely *which* principles are of the latter sort becomes the pressing problem for Descartes. In particular instances, he usually resolves the matter on the basis of being able to show a relationship of coherence and co-implication with his own methodology and his basic metaphysical propositions concerning the Cogito, God, and the more general meanings of nature. By submitting our natural beliefs to philosophical reconstruction within an order of reasoning responsive to the rigorous demands of the natural light or wisdom-seeking self, we can make a proper discernment of the rational status of such beliefs.

A prime instance of this discriminating process occurs in the vexed question of *finality*, when it is construed anthropomorphically to mean that the physical world is ordered solely or even primarily toward man's own welfare. Although in Descartes' time this belief was strong, pervasive, and apparently quite spontaneous, he subjected it to careful analysis in order to uncover the several layers of its epistemological structure. His findings indicated that, far from being a primary natural belief, this attitude toward material nature was compounded of three distinct factors: (a) a particular view of God, (b) a knowledge claim, and (c) a moral-religious suasion. Descartes argued that there is no necessary relationship between these three components (as there is between divine veracity and human inevitability of belief in the existing sensible world), and no sound basis in either metaphysics or the theory of nature for a contingent anthropomorphic union of them. The thrust of his critique of finalistic thinking was to neutralize its first two factors and reinstate the third one, but clearly on a new footing that entails no strict knowledge about nature.

(a) The issue does not concern whether God creates the universe through the operation of his intellect, will, and power. Descartes has his own independent grounds for accepting this traditional theistic position and thus for admitting that God is the purposive creator of nature. There are real ends for nature, in the sense that God creatively determines his significant purpose in producing and sustaining all natural realities. But the conventional notion of finality goes beyond this basic affirmation to hold that we men can somehow become privy to that purpose, can somehow convert the divine ends *for* nature into ends *found in* nature which fall within our comprehension and control.

This overclaim leads to the substitution of divination for a scientific and philosophic study of nature. Hence Descartes warns against conceiving of God as some sort of superman (*magnus aliquis homo*), who resembles us in proposing this or that future end to himself and then using these means or those to realize it. The infinite and creative God does not engage in this deliberating process, and hence does not operate through an instrumental sort of finality which might be open to human inspection. Thus it is a metaphysically evident implication of the Cartesian theory on the infinite power and intelligence of God that the human mind cannot strictly know the divine purpose operative in nature, although God may take the initiative to reveal some aspects of this purposiveness to our religious faith.

(b) Even on the contrary-to-fact condition that we could devise some intellectual means of ascertaining the divine purpose in creating the universe, Descartes would challenge its pertinence for modifying or enlarging his philosophical and scientific conception of nature. He discerns a twin flaw in every purported instance of human knowledge of the finality in nature: irrelevance and inflation.

Even though this or that divine end may be assigned for the processes in nature, the question arises of whether such assignment makes any methodological difference in our inquiries into natural processes. Descartes finds such finalistic interpretations singularly unenlightening, since they tell us nothing more about the mechanical structure and functioning of bodies and cannot even be related definitely to the mechanical modes of research. In addition to its explanatory irrelevance, finalistic thinking orders the relationships in nature too exclusively to a focal point of man. It rests upon too inflated an evaluation of man as constituting the sole, or even the primary, concern of God in the creative act. This overestimation of man neither provides a substitute for the Cartesian metaphysical account of the God-and-man relationship nor adds to the explanatory power of the Cartesian general meanings of nature.

There are adaptations and functional relations within the experienced world, and in this sense an *immanent* sort of purposiveness is present. But it can be analyzed sufficiently in terms of God as efficient cause (the first general meaning of nature) and in terms of the particulate components and mechanical laws and hypotheses (the other two general meanings of nature, supplemented by the more restricted modes of scientific inquiry). Our reliable scientific study of the world and our conception of nature are enriched by reference to these coordinates, but remain unaffected by this or that additional claim to know the transcendent divine purpose in creation.

(c) Why do men nevertheless persist in interpreting the natural process within some sort of divine purposive framework? Descartes' response depends upon distinguishing firmly between the rigorous reasoning required to obtain physical knowledge, and the humane persuasions allowed in matters of moral and religious conduct. He is willing to acknowledge a positive function for purposive thinking about nature, as long as this interpretation is directed toward the shaping of moral action and is not used as a premiss in constructing physical theories or our basic philosophy of nature.

> People are assured by religious faith and by metaphysical reasoning that God is the sole final cause of the universe, just as he is its sole efficient cause. And as for creatures, inasmuch as they subserve one another's ends reciprocally, each might ascribe to itself the privilege that whatever others subserve its ends are made "for its sake." . . . In ethics, indeed, it is an act of piety to say that God made everything for our sake, so that we may be the more impelled to thank him and be the more on fire with love of him. And in a sense this is true. For we can make *some* use of all things—at least we can employ our mind in contemplating them and in admiring God for his wonderful works. But it is by no means probable that all things were made for our sake, in the sense that they have no other use. In physical theory this supposition would be wholly ridiculous and absurd.[37]

[37] *Principles of Philosophy*, Bk. III, sect. 4, and *Letter of June 6, 1647*, to Chanut (AG, pp. 222–223; K, p. 222). The other texts used in the present analysis of the nature-and-finality problem are: *Meditations*, Bk. IV (Lafleur, p. 111); *Replies to Objections*, No. 5 (HR, vol. II, p. 223); *Entretien avec Burman*, ed. by C. Adam, *op. cit.*, pp. 46, 90; *Letters of August 1641, and September 15, 1645*, to Adversaries and to Elizabeth (K, pp. 117–118, 171–174). J. Laporte, *Le Rationalisme de Descartes* (2nd ed., Paris, 1950), pp. 355–361, observes that a properly founded judgment about the finality of the entire universe would have to depend on God's revelation of his own creative ordering of the whole assemblage of beings, activities, and relationships in nature—a total comprehension which surpasses in principle the claims made by Descartes for the philosophical understanding of nature. There

In this passage, Descartes sets off the general affirmation that God himself is the goal and significant purpose of all the processes in nature from the more particularized assertion that this or that adaptation within the universe discloses a special divine intention in creating things.

The latter view is not ruled out entirely, but it is surrounded by several epistomological restrictions. First, *any* entity capable of reflecting upon the instrumental relationships in the world, insofar as they serve its own well-being, could relate such relationships to the divine purposive activity. Any minded being, and not just man, could then claim that the other agencies were made by God for its own sake. Second, this reflective capacity is indeed found in man and does lead him to make reverent responses. To this extent, the rest of the universe can be regarded as serving to enhance man's religious attitude.

In the third place, however, this is a restricted moral-religious use of the relatedness of other natural realities to man. It does not imply that the significance of these realities is founded solely or primarily upon their enhancing relationship to human piety, since there are other dynamic interactions not subserving this use. And lastly, the moral-religious use of finalistic thinking cannot be exploited to increase our physical mode of knowledge or to modify our philosophically established meanings of nature.

Even under these qualifying conditions, Descartes detects a danger latent in this benevolent exegesis of our finalistic mode of thinking. For we may still try silently to erode the difference in principle between *noetic explanatory* requirements, holding for physical science and the philosophy of nature, and *moral dispositional* requirements for evoking our practical responses of gratitude and love of God. As a counterbalance operating from the practical side, therefore, he gives prominence to the further point that the religious attitude itself requires us to overcome homocentric egoism about the cosmos and the manner in which it manifests God's concern for man.

As a salutary corrective, Descartes commends Nicholas of Cusa and current astronomers for proposing the conception of an indefinitely large universe, which perhaps contains numberless other worlds. This view gives us a more adequate image of the divine immensity and creative power. Likewise, it furnishes scientific therapy against a foolishly homocentric appeal to purposive notions. Granted

---

is room in Cartesian philosophy of nature for mystery and unending research concerning nature and man, precisely because that theory of nature does not permit itself to be overwhelmed by finality claims.

that there may be other worlds following the universal mechanical laws and developing under conditions somewhat similar to ours, we can infer that they may well be peopled by beings analogous to men or even by far more intelligent creatures than ourselves.

On this likelihood, we must refrain from thinking that God has produced the universe solely or even primarily for the welfare of the human mode of intelligence. Thus a well-formed religious interpretation of nature should cautiously confine itself to arousing sentiments of love for the creative God and of mutual sympathy and help among all forms of rational creatures. It has no need to exaggerate by claiming an absolute supremacy for the values of man and his earth, or by centering the entire dynamic order of nature upon the satisfaction of human needs.

There are two last objections—respectively raised by humanism and religion—against the tempering of the finalistic outlook on nature by scientific and philosophic considerations. From the Renaissance humanistic perspective, the unique value of the human person seems to be undermined by the Copernican reversal of the earth's place in the universe and especially by the likelihood of other worlds and other minded beings. Descartes faces up to this difficulty in a thoughtful letter addressed to his friend, Ambassador Chanut.

> After believing for a long time that man has great privileges above other creatures, it looks as though we lose them all when we have occasion to change our view [on the relation between man and his earth to the rest of the universe being opened up by astronomical research and mechanical laws of nature]. I must distinguish between those advantages which can be diminished through others' enjoying similar ones, and those which cannot thus be diminished. . . . Now the goods that may exist in all the intelligent creatures of an indefinitely great world belong to this [latter] class: they do not diminish those that we ourselves possess. On the contrary, if we love God and for his sake unite ourselves in will to all that he has created, then the more grandeur, nobility, and perfection we conceive things to have, the more highly we esteem ourselves, as parts of a whole that is a greater work. And the more grounds we have to praise God for the immensity of his creation.[38]

[38] *Letter of June 6, 1647*, to Chanut (K, pp. 223–234). In his *Letter of August 18, 1645*, to Elizabeth (*Correspondance*, vol. VI, pp. 286–287), Descartes curtly reduces the Stoic maxim *follow nature* to a nonphysical, reflective moral meaning of nature. It enjoins us to seek that order of things established by God's will and conveyed to human reason, for our achievement of beatitude and wisdom. This he contrasts with every uncritical identification of the moral norm of nature with cultural mores and individual habituations. The Cartesian project of synthesizing science and philosophy, morality and religion, is examined by R. Lefèvre, *L'Humanisme de Descartes* (Paris, 1957); J. Combès, *Le Dessein de la sagesse cartésienne* (Paris, 1960); and J. Collins, *The Lure of Wisdom* (Milwaukee, 1962), pp. 40–122.

Thus man's dignity and religiousness are reinterpreted, but not annihilated, within the new perspective of nature as involving perhaps many worlds and many forms of intelligence.

The letter to Chanut is firm about three points in the Cartesian plan to reconstitute humanistic values within a scientific and mechanistic approach to nature. First the anxieties of the humanists refer to the moral dispositional order of man's self-evaluation and his worship of God. Although important, this order of interests cannot be permitted to shake the noetic foundations of the new conception of nature and man's function therein. Second, the traditional contrast between viewing man as somehow a whole and then as a part of the greater universe should be reconsidered, in order to bring out the humanistic implications of the partitive study of man. There is humane significance in seeing man as a part of a whole, that is, as a sharer in the greater work of the universe and a participator in a wider community of interpersonal values. Men are not diminished but enriched when personal values can be mutually cultivated with other intelligent agents, inhabiting other worlds. Lastly, Descartes notes that participation in such a broader community of knowledge and virtue liberates men from egoistic and earth-bound preoccupations, opening up the possibility of a magnanimity of spirit unknown to previous generations. The basic humanistic function of his new conception of nature is to open up the human self to the intellectual and moral values of the widening and cooperating universe, as seen in the Cartesian integration of reflective metaphysics and mechanistic science.

On the theistic/religious side, the remaining problem is that our mechanical analysis and technological control over nature seem to strip natural reality of all its mystery and wonder. Descartes does announce triumphantly, at the outset of his *Meteorology*, that the mechanistic explanation removes the wonder from the things we see in the universe.[39] He goes on to specify more carefully, however, that

---

[39] *Meteorology*, pts. 1 and 10 (Olscamp, pp. 263, 361). But exploration and steady technological advances give us a well grounded hope in human progress, and thus invest the study of nature with a justifiable emotive significance. "Carrying our sight much farther than the imagination of our fathers was used to going, they [the recently discovered telescopic lenses] seem to have opened the way for us to attain a knowledge of nature much greater and more perfect than our fathers had." *Optics*, pt. 1 (Olscamp, p. 65). The seventeenth-century shift in attitude towards nature, as a reality which we can strenuously explore, experiment with, and technologically mold to human purposes, is analyzed by R. Lenoble, "L'Évolution de l'idée de 'nature' du XVIe au XVIIIe siècle," *Revue de Métaphysique et de Morale*, vol. 58 (1953), pp. 108–129, and more at length in Lenoble's *Esquisse d'une histoire de l'idée de nature* (Paris, 1969), pt. 2.

his theory of nature is intended to eliminate one defective marvelling attitude based on two things: ignorance of the particulate structure and mechanical laws involved in natural events, and a sacral over-evaluation of nature as our mothering goddess. Yet the same scientific and philosophical knowledge of nature which erases a wonderment of this sort is also instrumental in arousing a different emotive response of reverence.

There is a strong basis for maintaining a sense of awe, hope, and respect under modern scientific and philosophic conditions. This more complex sentiment rests upon a fivefold weave of factors in the Cartesian approach to nature: acknowledgment of the unbounded vastness of the natural reality under investigation; methodically sustained awareness of the limits of the human mind which does the investigating; increasing hope in the power of technology to extend the domain of human presence and control in the universe; respect for every sort of natural being, down to the smallest ant; and consequently a more vivid appreciation, on the part of believers, of the way that God transcends the entire universe and its laws and also shows his intimate presence there. However it be renamed, then, the sense of wonder and mystery is not banished from man's awareness of nature, but is cleansed and put on a more reliable basis.

From the early stages in the formation of his theory of nature right down to its most mature development, Descartes has the teleological plan of integrating his philosophy of nature with these critically revised humanistic and religious values. The wisdom about nature toward which he strives is one that binds physical science with metaphysics and ethics. In working toward this contextual understanding of nature, he does more than offer a concrete realization of his own conception of how philosophy fulfills mankind's incessant drive toward wisdom, that is, toward combining an understanding and appreciation of nature with its technological reshaping and moral use. He also serves as an educator for the modern world in the making. He counsels humanists and believers not to remain suspicious and estranged from the new conception of nature, but rather to accept it as a creative gain and contribute to its further development. Descartes' philosophy of nature contains the practical imperative that the actual men that populate this our world—and not merely the fabular men of his preparatory tale—must increase the growth of knowledge, moral freedom, and religious awareness within the reality of nature.

Saint Louis University

References

1. DESCARTES: TEXTS AND TRANSLATIONS

Correspondance de Descartes, edited by C. Adam and G. Milhaud, 8 vols. (Paris, Presses Universitaires, 1936–63).

Entretien avec Burman, edited by C. Adam (Paris, Boivin, 1937).

Oeuvres philosophiques de Descartes, edited by Ferdinand Alquié, Vols. 1 and 2 (Paris, Garnier, 1963–67).

Principia Philosophiae, in *Oeuvres de Descartes*, edited by C. Adam and P. Tannery, Vol. VIII–1, new edition (Paris, Vrin, 1964).

Discourse on Method, Optics, Geometry, and Meteorology, translated by P. J. Olscamp (Indianapolis, Bobbs-Merrill, 1965).

Philosophical Essays: Discourse on Method; Meditations; Rules for the Direction of the Mind, translated by L. J. Lafleur (Indianapolis, Bobbs-Merrill, 1964).

Philosophical Letters, translated and edited by Anthony Kenny (New York, Oxford University Press, 1970).

Philosophical Writings, translated by E. Anscombe and P. T. Geach (New York, Nelson, 1954).

The Philosophical Works of Descartes, translated by E. S. Haldane and G. R. T. Ross, 2 vols. (New York, Cambridge University Press, 1967).

2. OTHER BOOKS

J.-L. Allard, *Le Mathématisme de Descartes* (Ottawa, Éditions de l'Université d'Ottawa, 1963).

A. G. Balz, *Descartes and the Modern Mind* (New Haven, Yale University Press, 1952).

L. J. Beck, *The Method of Descartes* (Oxford, Clarendon Press, 1952).

—— *The Metaphysics of Descartes* (New York, Oxford University Press, 1965).

Yvon Belaval, *Leibniz, critique de Descartes* (Paris, Gallimard, 1960).

Frederick Broadie, *An Approach to Descartes' 'Meditations'* (New York, Oxford University Press, 1970).

Gerd Buchdahl, *Metaphysics and the Philosophy of Science: The Classical Origins, Descartes to Kant* (Cambridge, Mass., M.I.T. Press, 1969).

M.-D. Chenu, *Nature, Man, and Society in the Twelfth Century* (Chicago, University of Chicago Press, 1968).

James Collins, *The Lure of Wisdom* (Milwaukee, Marquette University Press, 1962).

Joseph Combès, *Le Dessein de la sagesse cartésienne* (Paris, Vitte, 1960).

J. A. Comenius, *The Great Didactic of John Amos Comenius*, translated by M. W. Keating, 2 vols. (London, Black, 1921).

Elie Denissoff, *Descartes, premier théoricien de la physique mathématique* (Louvain, Nauwelaerts, 1970).

E. J. Dijksterhuis, *The Mechanization of the World Picture* (Oxford, Clarendon Press, 1961).

A. B. Gibson, *The Philosophy of Descartes* (London, Methuen, 1932).

Étienne Gilson, *Index scolastico-cartésien* (Paris, Alcan, 1912).

—— *René Descartes: Discours de la méthode, texte et commentaire*, 2nd edition (Paris, Vrin, 1947).

Henri Gouhier, *La Pensée métaphysique de Descartes* (Paris, Vrin, 1962).

Marjorie Grene, *The Knower and the Known* (New York, Basic Books, 1966).

Martial Guéroult, *Descartes selon l'ordre de raisons*, 2 vols. (Paris, Aubier, 1953).

F. C. Haber, *The Age of the World: Moses to Darwin* (Baltimore, Johns Hopkins Press, 1959).

P. P. Hallie, *The Scar of Montaigne* (Middletown, Wesleyan University Press, 1966).

M. B. Hesse, *Models and Analogies in Science* (Notre Dame, University of Notre Dame Press, 1966).

Alexandre Koyré, *Newtonian Studies* (Cambridge, Harvard University Press, 1965).

Jean Laporte, *Le Rationalisme de Descartes*, 2nd edition (Paris, Presses Universitaires, 1950).

Roger Lefèvre, *L'Humanisme de Descartes* (Paris, Presses Universitaires, 1957).

Robert Lenoble, *Mersenne, ou la naissance du mécanisme* (Paris, Vrin, 1943).

—— *Esquisse d'une histoire de l'idée de nature* (Paris, Michel, 1969).

W. von Leyden, *Seventeenth-Century Metaphysics* (New York, Barnes and Noble, 1968).

Robert McRae, *The Problem of the Unity of the Sciences: Bacon to Kant* (Toronto, University of Toronto Press, 1961).

Michel de Montaigne, *The Complete Works of Montaigne*, translated by D. M. Frame (Stanford, Stanford University Press, 1957).

Richard Popkin, *The History of Scepticism from Erasmus to Descartes*, revised edition (New York, Harper, 1968).

J. F. Scott, *The Scientific Work of René Descartes (1596–1650)* (London, Taylor and Francis, 1952).

Sextus Empiricus, *Scepticism, Man, and God: Selections from the Major Writings of Sextus Empiricus*, translated by S. G. Etheridge and edited by P. P. Hallie (Middletown, Wesleyan University Press, 1964).

N. K. Smith, *New Studies in the Philosophy of Descartes* (London, Macmillan, 1952).

J. R. Vrooman, *René Descartes, A Biography* (New York, Putnam, 1970).

Jules Vuillemin, *Mathématiques et métaphysique chez Descartes* (Paris, Presses Universitaires, 1960).

3. ARTICLES AND ESSAYS

R. J. Blackwell, "Descartes' Laws of Motion," *Isis*, vol. 57 (1966), pp. 220–234.

Émile Bréhier, "The Creation of the Eternal Truths in Descartes's System," in *Descartes*, edited by Willis Doney (New York, Doubleday, 1967), pp. 192–208.

Gerd Buchdahl, "Descartes's Anticipation of a 'Logic of Discovery'," in *Scientific Change*, edited by A. C. Crombie (New York, Basic Books, 1963), pp. 399–417.

P. H. J. Hoenen, "Descartes's Mechanicism," in *Descartes*, edited by Willis Doney (New York, Doubleday, 1967), pp. 353–368.

Julian Jaynes, "The Problem of Animate Motion in the Seventeenth Century," *Journal of the History of Ideas*, vol. 31 (1970), pp. 219–234.

Robert Lenoble, "L'Évolution de l'idée de 'nature' du XVIe au XVIIIe siècle," *Revue de Métaphysique et de Morale*, vol. 58 (1953), pp. 108–129.

Ernan McMullin, "Philosophies of Nature," *The New Scholasticism*, vol. 43 (1969), pp. 29–74.

T. P. McTighe, "Galileo's 'Platonism': A Reconsideration," in *Galileo: Man of Science*, edited by Ernan McMullin (New York, Basic Books, 1968), pp. 365–387.

Patricia Reif, "The Textbook Tradition in Natural Philosophy, 1600–1650," *Journal of the History of Ideas*, vol. 30 (1969), pp. 17–32.

INDEX

Descartes is here indexed by his topics and italicized writings.